Microcomputers in physiology

a practical approach

TITLES PUBLISHED IN
THE PRACTICAL APPROACH SERIES

Series editors:
Dr D Rickwood
Department of Biology, University of Essex
Wivenhoe Park, Colchester, Essex CO4 3SQ, UK
Dr B D Hames
Department of Biochemistry, University of Leeds
Leeds LS2 9JT, UK

- Affinity chromatography
- Animal cell culture
- Antibodies
- Biochemical toxicology
- Biological membranes
- Carbohydrate analysis
- Centrifugation (2nd Edition)
- DNA cloning
- Drosophila
- Electron microscopy in molecular biology
- Gel electrophoresis of nucleic acids
- Gel electrophoresis of proteins
- Genome analysis
- HPLC of small molecules
- Human cytogenetics
- Human genetic diseases
- Immobilised cells and enzymes
- Iodinated density gradient media
- Lymphocytes
- Lymphokines and interferons
- Mammalian development
- Microcomputers in biology
- Microcomputers in physiology
- Mitochondria
- Mutagenicity testing
- Neurochemistry
- Nucleic acid and protein sequence analysis
- Nucleic acid hybridisation
- Oligonucleotide synthesis
- Photosynthesis: energy transduction
- Plant cell culture
- Plasmids
- Prostaglandins and related substances
- Spectrophotometry and spectrofluorimetry
- Steroid hormones
- Teratocarcinomas and embryonic stem cells
- Transcription and translation
- Virology
- Yeast

Microcomputers in physiology

a practical approach

Edited by
P J Fraser
Department of Zoology, University of Aberdeen, Aberdeen
AB9 2TN, UK

OXFORD · WASHINGTON DC

IRL Press
Eynsham
Oxford
England

© IRL Press Limited 1988

First Published 1988

All rights reserved by the publisher. No part of this book may be reproduced or transmitted in any form by any means, electronic or mechanical, including photocopying, recording or any information storage and retrieval system, without permission in writing from the publisher.

British Library Cataloguing in Publication Data

Microcomputers in physiology.
  1. Physiology. Applications of microcomputer systems
  I. Fraser, P.J.   II. Series
  574.1′028′5416

Library of Congress Cataloging in Publication Data

Microcomputers in physiology.
  (The Practical approach series).
  Includes index.
  1. Physiology—Data processing.   2. Microcomputers.
  I. Fraser, P. J.   II. Series.
  QP33.6.D38M53 1988   599′.01′0285   88-13312
  ISBN 1 85221 129 6 (hardbound)
  ISBN 1 85221 130 X (softbound)

Typeset by Infotype and printed by Information Printing Ltd, Oxford, England.

Preface

Microcomputers have become familiar items in all physiological laboratories. They perform a number of roles, replacing or complementing the chart recorder or oscilloscope, the scientist with his ruler, the statistician's specialized calculator and the artist's handproduced graphs. In the course of a decade, they have changed from relatively limited controllers to extremely powerful devices rivalling the mainframe computers which were in operation when microcomputers first appeared.

Coming after the Practical Approach Series volume *Microcomputers in Biology: A Practical Approach*, this book has less need to emphasize the concept of a microcomputer as an item of laboratory equipment, or explain introductory jargon. Microcomputers are extremely versatile items, a recipe book approach to using them is inappropriate. Microcomputer users suffer not from a paucity of bench manual material, but from a considerable surfeit. A recipe explains and dictates. Microcomputing recipes are most useful if they provide wide explanations in the course of limited dictated example. The inexperienced reader needs simple wide explanation. The experienced reader is seeking realistic appraisal of microcomputing limits in the context of extensively worked examples. In view of the rapid evolution and expansion of the microcomputer, its limits and its uses, scientific personnel are split in terms of knowledge and experience in an unparalleled way. The split often reverses the normal hierarchy based on experience increasing with age. The split will be healed as more and more gain productive hands-on experience. Use of computers in one context, for example word processing, usually leads to more relaxed adoption of microcomputers in other contexts.

The rapid development of microcomputers from cheap but rather limited data loggers, to versatile, easy to use devices which are not only cheaper than other forms of data logging, but considerably better has been a significant gain for the scientific community. What is just now being realized is the potential for new experiments, new approaches and higher levels of measurement and working practice. The aim of this book has been to give simple accounts of selected systems which have been extensively developed by physiologists dedicated to developing the new microprocessor-based technology. In the course of this it is hoped to provide a practical introduction to a wide range of techniques.

In the first chapter, a brief and simplified introduction to computers and computer jargon is intended to complement existing introductory accounts. In this chapter and the following one, various interfacing ideas, solutions and problems relevant to the neurophysiological laboratory (and hence to other branches of physiology which require less fast data acquisition) are presented. Some software is presented in appendices to the chapters, more is available from the authors. The next two chapters concentrate on the anatomical side, with accounts on manual and automatic image acquisition and manipulation. The possibilities in the future for automatic image interpretation are immense and exciting. Cardiac data logging has been one of the most extensively developed fields. Locomotion studies form a good example of the expansion of data acquisition possible under computer control. In Chapter 7, the development of computer-based stimulators to help paraplegics is a good example of the direct application to modern medical practice of physiological techniques which were not possible before the advent of the microprocessor. As a slight departure from the normal approach of this series, the application is placed in the context of a review of the uses of electrical stimulation. The potential of direct links between computer and nervous system or muscle systems (heart and locomotor muscles) defies description. Finally, the all-encompassing subject of computer-aided teaching is particularly significant to the physiologist, since the development of realistic simulation systems replacing many student experiments on animals promises the removal of what has been a significant block on a broad practical approach to learning physiology.

It is hoped that the examples selected in this book act as pointers to the future use of the microcomputer in the physiological laboratory, not only as a polished instrument which is a useful alternative to existing equipment, but as an instrument which opens up new approaches not possible in any way before.

P.J.Fraser

Contributors

J.J.Capowski
Department of Physiology, Campus Box 7545, Medical Sciences Research Building, University of North Carolina, Chapel Hill, NC 27514, USA

J.Dempster
Department of Physiology and Pharmacology, University of Strathclyde, Royal College, 204 George Street, Glasgow G1 1XW, UK

P.J.Fraser
Zoology Department, Aberdeen University, Tillydrone Avenue, Aberdeen AB9 2TN, UK

G.V.Moore
Department of Computer Science, University of Manchester, Oxford Road, Manchester M13 9PL, UK

J.S.Petrofsky
University of California, Irvine, Veterans Administration, Long Beach, 23832 Rockfield, Suite 289, Lake Forest, CA 92630, USA

M.R.Pierrynowski
School of Physical and Health Education, University of Toronto, 320 Huron Street, Toronto, M5S 1A1, Canada

J.E.Randall
Indiana University School of Medicine, Medical Sciences, Myers Hall, Bloomington, IN 47405, USA

E.Skordalakis
Department of Electrical Engineering, Division of Computer Science, National Technical University of Athens, 15773 Zographou, Athens, Greece

P.Trahanias
NRCPS 'Democritos', Institute of Information and Telecommunications, Aghia Paraskevi, Athens 15310, Greece

Acknowledgements

Trademarks/Owners

Xenix, MS-DOS, QuickBASIC, QuickC and Microsoft/Microsoft Corp.
IBM/IBM Inc.
Turbo BASIC, Turbo Pascal/Borland International Inc.
Apple, Macintosh/Apple Computers Inc.

All other brand and product names are trademarks of their respective parent companies.

Contents

ABBREVIATIONS	xv

1. INTRODUCTION   1
P.J.Fraser

History of Computers in Physiology	1
Levels of Use and Understanding	2
Introduction to Machine Code Programming	4
Introducing characters, data and memory in microcomputers	4
Data, programs and ASCII codes	8
Machine code programming	12
Low Budget Application of Microcomputers to Equilibrium Research	17
Interface requirements	17
Servomotors and D/A converters	17
Timing and spike counting with CTC	17
A pointer to machine code required	19
Using an ADC to monitor muscle potentials	21
Using a higher level language to control a machine code module	24
Good and bad programs	25
Conclusion	26
References	26
Appendix: Program Listings	27

2. COMPUTER ANALYSIS OF ELECTROPHYSIOLOGICAL SIGNALS   51
J.Dempster

Introduction	51
Computer Hardware for Digitizing Analogue Signals	52
The laboratory interface unit	52
Data transfer from interface to computer	55
Signal Conditioning	57
Analogue signal conditioning	57
Filtering signals	57
Synchronization trigger signal	60
A simple signal conditioning circuit	60
Computer Hardware	61
The IBM PC and compatible computers	61
The MS-DOS operating system	61
Data storage	62
Display of graphics	63
Hard copy	64

A Software Package for Analysing Electrophysiological Data	65
Electrophysiological analysis software	65
Inspection and processing records	69
Automatic measurement of signal records	73
Curve fitting	75
Analysis of patch clamp recordings	77
Channel open/close transition detection	79
Open and closed time histograms	82
Noise analysis	83
Command voltage pulse generation	86
'On-line' versus tape	87
Pulse code modulation recording	88
Development of Scientific Software	88
How to obtain software	88
Software design goals	89
Programming languages	89
Graphics virtual device interfaces	90
Some commercially available software	91
Acknowledgements	92
References	93

3. ANATOMICAL MEASUREMENT AND ANALYSIS — 95
J.J.Capowski

Introduction: Why Apply Computers to Neuroanatomy?	95
The combination of human and computer	95
A semiautomatic system for collecting data	95
Semiautomatic Tracing with a Cursor	96
Techniques of tablet tracing	96
Feedback is important	99
Elementary Two-dimensional Tablet Measurements	100
Marking the locations of cells and other structures	100
Counting features	101
Calculation of two-dimensional structural parameters	102
Population Calculations from these Parameters	104
Histograms	105
Distributions	105
Density measurements	105
Three-dimensional Computer Graphics	106
Three-dimensional Serial Section Reconstruction	110
Section alignment	111
Three-dimensional measurements	112
Neuron Reconstruction	112
Computer hardware for tracing neurons	112
The neuron tracing procedure	114

Merging of multiple section dendrites	115
Intermediate computations	116
Three-dimensional display of the neurons	118
Plots of the neuron structure	119
Statistical summaries of individual neurons	119
Population summaries	120
The Use of Video	121
Computer–television hardware	121
Cursor mixing using video	122
Image enhancement	123
Extracting numerical information from an image	125
Closing Comment	126
Acknowledgements	127

4. DIGITAL IMAGE PROCESSING AND ANALYSIS TECHNIQUES   129
G.V.Moore

Introduction	129
Basic principles of digital image manipulation	129
Image Manipulation	130
Image storage and representation	130
Colour images	131
Neighbourhoods	132
Connectivity	132
Basic Hardware Requirements	133
Image input devices	133
Video cameras	134
Other image input devices	135
Image display devices	136
Frame stores	136
Processors	137
Simple Image Processing Techniques	137
The grey-level histogram	137
Point transformations	139
Local Operators	141
Image smoothing	142
Edge detection	143
Implementation of local operators	144
Image Analysis Techniques	145
Freeman chain code	146
Feature measurements	147
A Simple Cell Analysis Scheme	148
Pre-processing	149
Segmentation	149
Feature selection	149

    Feature measurement   151
Summary   154
Further Reading   154
References   155

5. COMPUTERS IN CARDIAC RESEARCH   157
E.Skordalakis and P.Trahanias

Introduction   157
ECG Data Handling   157
    Electrocardiographic patterns   159
    ECG acquisition   160
    Signal conditioning   161
    ECG recognition and parameter measurement   162
    An illustrative example   173
    ECG interpretation and classification   174
Cardiac Image Handling   174
    Cardiac imaging techniques   174
    Cardiac image processing   175
References   176
Appendix   177

6. COMPUTER ANALYSIS OF LOCOMOTION   179
M.R.Pierrynowski

Introduction   179
Computer System and Locomotion Laboratory Equipment   180
    Input devices   180
    Output devices   182
Locomotion Data Analyses   183
    Input data   186
    Output data   188
Conclusion   191
References   191

7. ACTIVE PHYSICAL THERAPY—COMPUTERIZED REHABILITATION   193
J.S.Petrofsky

Introduction   193
Functional Electrical Stimulation   193
Exercise Devices Involving Computer-controlled Movement   198
Computer-controlled Movement   211
Acknowledgements   218
References   218

8. SIMULATION TECHNIQUES FOR TEACHING — 221
J.E. Randall

Introduction	221
Physiological Simulation	221
Why acquire programming skills?	223
Hardware	224
Architectures	224
The IBM-PC	225
Use in lecture halls	227
Software	229
Simulation languages	229
Drivers for text files	229
General-purpose languages	230
Using the Compiler	234
Linking compiled program modules	234
Compiling to executable code	234
Compiling to memory	235
Interactive Techniques	236
Technical suggestions	237
Utility subroutines	239
Graphics Subroutines	240
Subroutine for one graph	241
Subroutine for four graphs	242
User-defined Functions	243
Non-linear function generator	244
A simple teaching example	245
Large Mathematical Models	246
The comprehensive model	247
A simple example	248
Acid/base balance	250
Multiple graphs	251
Combining model with graphics by text files	252
Teaching Protocols	255
Objectives	255
Appropriateness	256
Environment	257
Acknowledgements	258
References	259
Note Added in Proof	259
Appendix: Program Listings	260

APPENDIX

Suppliers of Specialist Items	271

INDEX — 273

Abbreviations

ACh	acetylcholine
A/D	analogue-to-digital
ADC	analogue-to-digital converter
CAT	computer assisted tomography
CCD	charge coupled device
CGA	colour graphics adapter
CPS	characters per second
CPU	central processor unit
CRT	cathode ray tube
C/S	control/status
CT	computed X-ray tomography
CTC	counter/timer circuit
D/A	digital-to-analogue
DAC	digital-to-analogue converter
dB	decibel
DMA	direct memory access
DPI	dots per inch
ECG	electrocardiogram
EGA	enhanced graphics adapter
EMG	electromyograms
FES	functional electrical stimulation
FFT	fast Fourier transform
FR	frame of reference
GEM	Graphics Environment Manager
HPGL	Hewlett Packard Graphics Language
I/O	input/output
LCD	liquid crystal display
LED	light emitting diode
MCP	mean circulatory pressure
MEPC	miniature end-plate current
MEPP	miniature end-plate potential
MRI	magnetic resonance imaging
NLQ	near letter quality
PC	personal computer
pdf	probability density function
PIO	parallel I/O device
PLA	piecewise linear approximation
QB	QuickBASIC
SCOP	Simulation Control Program
S/H	sample and hold
TB	Turbo BASIC
TENS	transcutaneous electrical nerve stimulation
TTL	transistor−transistor−logic
VDI	virtual device interface

CHAPTER 1

Introduction

PETER J.FRASER

1. HISTORY OF COMPUTERS IN PHYSIOLOGY

During the development of the laboratory computer in the 1950s, applications to sensory physiology and information processing were amongst the first uses of the machines. Scientists at MIT's Lincoln Laboratory where the several billion dollar SAGE continental air defence system for the US Air Force was based, were the first to *'demonstrate how a digital computing device could be employed to control, record or analyse, with enormous flexibility and very great precision, a remarkable diversity of environmental events with respect to both spatial and temporal parameters, and even today we have hardly yet begun to exploit this truly ingenious insight'* (1). These words were written just before the microcomputer revolution which has brought vastly increased use of computer technology.

Cost separated laboratories into the 'haves' and the 'have nots'. Much was done on the early minicomputer systems (1,2). Many PDP 11 computer systems are still running and performing well after 10 years. The early systems although sparse in number were well supported in terms of computer programmers. Much special software was written in machine code, BASIC, which was slow and limited at that time and FORTRAN which was the main scientific language used in mainframe computers. At first, microcomputers came into physiological laboratories as home built microprocessor systems, or first and second generation microcomputers. These used tape recorders to store programs and data, had 1−4 K memory, and generally were used for some specialized interface task. Data was passed to a mainframe or mini computer for further analysis. This early phase in the development of microcomputer applications to physiology is poorly represented in the literature. The breakthrough was largely that of cost, and those involved were caught up in a spiral of rapidly developing technology which allowed exponential growth of memory capacity, processing speed and graphics handling ability.

Early microcomputers used their central microprocessors in a very direct manner so that interfacing was rather straightforward. Later microcomputers used facilities such as interrupts as part of the microcomputer architecture for controlling discs and other peripheral devices. It was now necessary to have a detailed knowledge of what had become a very complex machine to attach and program peripheral devices. In general specialized interfaces with A/D converters, D/A converters, counters, timers, digital input/output and relays for high current switching have evolved from the early microcomputer based experimental interfaces. These contain large memory buffers and communicate directly with well supported microcomputer architectures such as the IBM PC. Modern microcomputers make use of powerful 32-bit microprocessor chips, access

Introduction

reliable floppy discs, or hard discs, and have upwards of one Megabyte of memory allowing large programs to run which carry much of the prompting information necessary to run them. Yet these machines still cost less than the early development kits.

Languages such as BASIC, which was developed largely for teaching, have been expanded into well integrated powerful languages. BASIC was developed as the primary microcomputer language. Commands were added to control the appended facilities such as sound, colour graphics and analogue-to-digital conversion. Procedures and functions appeared and finally compilers brought the whole package up to the top end of the higher level language range (see Chapter 8). With larger memory capacities, and efficient disc stores available, most mainframe languages are now available on microcomputers, so there is a choice of FORTRAN, Pascal, C, FORTH, BCPL and others. Pascal is well structured, teaches excellent programming habits, and has a Standard form described, easing portability. Paradoxically, this has prevented evolution of the language with improvements in hardware. My first choice of language is now compiled BASIC. In the future, expert systems, fuzzy logic and a range of artificial intelligence techniques are set to dominate the next generation of microcomputers. Mice and windowing techniques are of advantage in controlling software and choosing options, although less used in scientific applications.

Microcomputers have been usefully applied by the physiologist in a large number of ways ranging from data logging, statistical analyses and graph plotting, to image storage, manipulation and quantification. Many of the early interface applications are now commercially manufactured, with extensive software support. The present generation of microcomputing physiologists are more concerned with handling the vast amount of computer assisted data becoming available. Due to the microcomputer revolution, the whole scientific community is split into different levels of understanding of computers. In this chapter I attempt a brief introduction to computing jargon which I know causes difficulty, but which will crop up quickly in any application. I also present some examples of interfacing tasks developed from first principles on an early Z80 microcomputer, the Nascom. The intention is largely to give an understanding of the sorts of ways laboratory interfaces work, rather than to urge the reader to revert to an 8-bit microcomputer (although Z80 machines are still prominent in the microcomputer world).

2. LEVELS OF USE AND UNDERSTANDING

The Practical Approach Series which includes this volume intends use of each book as a bench manual for novices for a particular range of related techniques. One of the problems facing the novice computer user is not the lack of a bench manual, but a surplus of instructions. Looking at the set of manufacturers manuals I have for the microcomputers in my laboratory, and the manuals for the small (strictly finance limited) set of software, the most convenient summary is a width measurement of A4 material of just short of 1 m. It must be recognized that for even the experienced microcomputer user, buying a new machine involves about 3 months learning/familiarization of new commands and systems and slightly different old commands. Once familiarization is achieved, it is still necessary to refer often to the manuals. The best and only approach to using microcomputers is to use them and use them often.

The physiologist requires knowledge of the range and scope of microcomputers and their interfaces. He needs to know whether there exists software suitable for his particular task, or if the software is not available how he could produce it or have it produced. Given interfaces and software he needs to know how to use it most efficiently and how to handle and process data. Instructions for use could range from a statement of which key to press to start the process for a largely automatic task to a series of tasks involving selection of particular interfaces; wiring them up appropriately; writing machine code interface software or modifying software already written for a broadly similar task; linking machine code modules to a higher level language package for convenient manipulation and parameter passing; production of data files; data filtering; graph plotting and finally report writing.

Microcomputers are cheap. Different microcomputers are good at and may conveniently be set up to do different tasks. It is often most convenient to have several machines each specialized for different tasks. It is no good having a machine set up as a data logger on an experimental setup if the same machine is required at the same time for word processing or running statistical software. Two obstacles hinder efficient use of several microcomputers. First there is a built-in resistance amongst administrators who allocate money for resources to allowing multiple copies of equipment for tasks which could conceivably be done by a single machine. Do not underestimate the task of physically moving a microcomputer setup with monitor, printer, interface boxes, disc drives and communications links. Do not underestimate the task of setting switches from one configuration to another. Proper testing of the microcomputer in its new environment and configuration could take hours or days. Also, there is usually a need to pass data and special purpose programs between machines. The easiest way to do this is by having compatible systems and move floppy discs between machines. This is simplest where machines are identical.

Disc drives are better constructed than they used to be. I have seen 'identical' systems which would allow discs to be read from machine A to machine B but not *vice versa* because of slight differences in the settings of the recording heads on the disc drives. This sort of problem is less common now. Where machines have evolved for several years as in BBC B and IBM PC configurations, there may be problems moving discs between early disc systems and later ones.

Where it is not possible to move discs, it is often possible to link machines through serial or parallel ports. KERMIT is a serial transfer program which will allow files to be moved from one machine running KERMIT to another machine running KERMIT. Because there is extensive error checking, transfer is slow, but command files can be set up to allow automatic transfer of large numbers of files at a time convenient to the user such as overnight. KERMIT versions exist for most microcomputers and mainframe computers allowing transfer of files all over the country. If possible, KERMIT makes the transferred data usable to the recipient machine. KERMIT versions are usually available for the cost of the carrying medium (mostly floppy disc). The system was devised at the Center for Computing Activities at the University of Columbia in New York. CUCCA retain copyright on KERMIT, but have published full information on it and permit anyone to implement it on their own machines, provided this is not done for commercial purposes. In the UK, the scheme is coordinated through the University of Lancaster. Network configurations which allow several machines to share facilities

Introduction

such as disc drives are available for many well supported microcomputers. These allow file transfer, but may impose some limits on their constituent computers.

Choosing a microcomputer system is governed by what is currently available (who could predict the computing power available in 2 years time?); what is well supported in terms of choice of software and interfaces; what fits with local policy; what has local technical backup; what can be afforded. There are trade-offs often between getting the most powerful machine available and getting support for it since support takes $2-4$ years to reach a useful level. Where tasks can be split, it may be more efficient to have two machines than one which is twice as fast and twice as expensive. It is as well to also consider the space required for one or several microcomputers. Although portable machines are more expensive, they might be of advantage where space on equipment racks is limited. Microcomputers give surprisingly little trouble from the interference point of view, but it is necessary to have adequate screening on video and transfer leads.

3. INTRODUCTION TO MACHINE CODE PROGRAMMING

3.1 Introducing characters, data and memory in microcomputers

This section is for those with no understanding of what goes on in a computer. It is filled with oversimplifying statements to try to give a first level of understanding, but should define bit, byte, binary, hexadecimal, RAM, ROM, K and M. Beginners are urged to scan the vast array of books and magazines now on the market. If in doubt ask your young children! Good introductory accounts more relevant to the physiologist now exist $(3-5)$. Chapters 1 and 2 of reference 3 and Chapter 2 of reference 4 are essential reading. There are hundreds of books now available. When I first built and used microcomputers only a decade ago, I had to teach myself from the chip maker's manual.

3.1.1 Binary numbers and data bytes

Let us start with the idea of the computer handling and storing numbers in memory and elsewhere. First a little on memory terms and the byte. The term 'K' is now common parlance, as in 'My computer has 500 K'. This refers to memory capacity and to understand its exact meaning it is necessary to know what a binary number is.

(i) *Binary numbers*. Just as the decimal (base 10) number 123 means $1 \times 10^2 + 2 \times 10^1 + 3 \times 10^0$ or $100 + 20 + 3$ with digits chosen from the ten used in the English language (0, 1, 2, 3, 4, 5, 6, 7, 8 and 9), the binary number 101 means decimal 5. We treat the number in the same way as usual, but in binary, there are only two digits (0 and 1) and the base is 2. Binary 101 means $1 \times 2^2 + 0 \times 2^1 + 1 \times 2^0$ or $4 + 0 + 1$. A three digit binary number can take the maximum value binary 111 or decimal 7. To calculate the maximum value for a given number of digits of a number to a particular base, calculate the base to the power of the number of digits and subtract 1. Each BInary digiT or *bit* is represented in the microcomputer as a high (usually ~ 5 V) value, often known as '1' or a low (~ 0 V) level, often known as '0'. If you have an eight digit binary number, then the largest value it can take is binary 11111111 or $2 \times 10^8 - 1$, which is decimal 255. There are 256 different values possible $(0-255)$. Physically you can think of eight conducting wires switched high to 5 V or low to 0 V. There are two common exceptions to the 0 V, 5 V signalling system. Special memory

4

chips called ROMS or EPROMS [see Section 3.1.1(iii)] often have 12 V power lines to them which must be kept separate from the 5 V TTL (transistor−transistor−logic) circuitry. Serial RS232 communication lines commonly use +12 V and −12 V levels for logical 1 and logical 0.

(ii) *Memory addresses*. Data are stored in memory as one or more 8-bit numbers (bytes). The access and storage of such data is performed by the microprocessor or central processor unit (CPU) in the heart of the microcomputer. Early microprocessors which go under abrupt names such as Z80 or 6502 deal with their numbers in chunks of 8 bits. The address number is a 16-bit number, so that one of these 8-bit microprocessors can access a total of 65536 memory locations (addresses 0−65535). Despite the ability to address more than 65 thousand memory locations, an 8-bit microprocessor addressing its full complement of memory is said to be addressing 64 K of RAM. The unit 1 K (Kilobyte or kbyte) is a 'binary' thousand [2^{10} or 1024 units (bytes) of memory]. It is inappropriate for modern microcomputers which allow 20- or 24-bit addresses with 1048576 [1 M (Megabyte or Mbyte)] or 16 777 216 bytes (16 M) possible.

(iii) *Memory types*. RAM is 'ordinary' memory which only retains its contents when powered up. When a microcomputer starts up, it begins to read instructions which are bytes stored in memory such as 62 0 or 211 4. These instructions mean little to us, but to the particular microprocessor, the combination of 0s and 1s or high and low levels act as combination switches to trigger certain operations. Of course, since RAM is wiped out on loss of power, after switch on, all memory will contain nothing meaningful. A special permanent memory is used to contain the instructions necessary for the starting sequence. This is physically in a special memory chip known as ROM or EPROM. Telling you that RAM stands for random access memory, and ROM stands for read only memory does not really help you understand their function. The internal structure of a ROM chip is similar to a very complex matrix type fuse board. During programming, individual fuses are blown or not to represent digital logic 1 or logic 0. Once blown they cannot be altered. Electrically Programmable ROM or EPROM is a form of permanent memory which can be erased by UV light treatment and reprogrammed on a special EPROM programming machine. The terms are worth knowing since many programs come in the form of an EPROM chip which plugs into a special socket on the microcomputer and are readily fitted by the amateur provided he remembers to switch off all power. Volatile RAM stores a number only for a short time. Once stored and left, the bit pattern will fairly quickly break down. It is necessary for the number in the memory store to be accessed and rewritten at short time intervals. A microprocessor is designed to maintain the data in dynamic RAM in this way and contains a special refresh register which increments through the memory addresses so that each byte of data is accessed and rewritten in turn. Apart from telling you a little more about RAM, it is necessary to know about the refresh register since it is often accessed to provide a *random number*. Accessed once in a program, this will be a true random number. However, since the refresh register is incremented on a strict time sequence, when accessed again in a program, the number obtained will be related to the first according to the relative times of access. This caveat must be borne in mind when utilizing some random number functions on a microcomputer. With modern low power consumption memory chips, power can be retained using a small battery built

Introduction

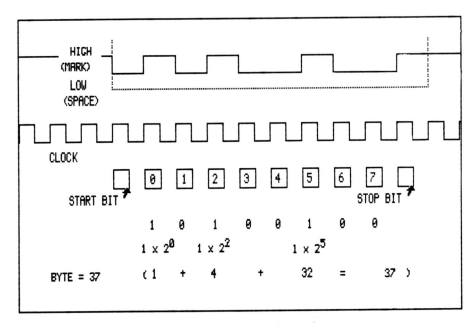

Figure 1. Serial transfer of an 8-bit byte. The number 37 decimal is 10100100 binary. To transmit the byte, the serial line (top trace) goes from high to low for one clock period. The state of the serial line for the next clock period represents bit 0, and so on for successive bits. The line finally goes high for one clock period (the stop bit).

in close to the ram chips for days or weeks. Such battery backed RAM can be removed from the machine and will retain the contents of its memory.

3.1.2 Data storage on magnetic media
Using battery backed RAM is not the only way the computer's bytes can be stored outside the machine. Magnetic media have a long life and can be accessed fast. Paper tape and punched cards are practically obsolete. Interestingly enough, but not strictly relevant to the physiologist, for long term data storage such as sealing data into time capsules, punched hole technology on a substrate more permanent than paper is preferred.

(i) *Serial data transfer.* Clearly a set of eight voltages on eight lines represents a byte of data. These voltages could be stored permanently on eight tracks of a magnetic tape and later played back. It is only necessary to play such a record for a very short time for it to be properly read by the computer. In practice, eight channel tape recorders are expensive, so a single recording channel is used. The voltages in each line are represented sequentially on a single channel. A special chip called a UART (universal asynchronous receive and transmit) chip is often used to do the necessary conversion. Thus the level of bit 0 is read and then recorded for a short time followed by that of bit 1, and so on until all 8 bits are recorded. Add start bits and stop bits so that the pattern of data is structured and you can orient to the eight data bits from any point in the data stream and you have the basis of serial data transfer (*Figure 1*).

To record data on an ordinary audio tape recorder, a frequency code is used whereby a logical 1 is represented by a short high frequency signal and a logical 0 as a short lower frequency signal. The key to decoding a serial stream of data is knowing how long the voltage representing each bit of the byte is switched on. Normally the process of recording or reading serial data is controlled by an oscillating square wave. This timing waveform which acts as and is known as a clock, gates access to each bit of the waveform. It is common to use one start bit and one stop bit, making 10 bits necessary to transfer a single 8-bit byte. Within a computer, the clock frequency or *baud rate* is controlled by an external baud rate generator chip which allows selection of a series of standard rates such as 110, 300, 1200, 4800, or 9600 baud. If a single start and stop bit is used, then 9600 baud transmits a maximum of 9600 bits/sec which corresponds to 960 bytes/sec.

(ii) *Discs*. The commonest magnetic storage device used at present is the floppy disc which is a thin plastic disc coated with magnetic medium and mounted in a cardboard (for 5.25 inch discs) or plastic envelope with a metal guard (for more advanced 3.5 inch discs). The disc spins round and a recording/playback head moves radially in short steps from the periphery to the centre. When the head is lowered close to the disc, data can be read or recorded. Data can be quickly accessed from any point on the disc. Compare this with a tape recorder where several hundred meters of tape might have to be wound in before a given point could be accessed. A hard disc or Winchester disc is similar in principle, but a non-removable carefully engineered disc is used with recording heads set so close to the disc surface that the particles in tobacco smoke would be sufficient to jam between the disc head and the disc. Such discs come in a sealed form to prevent such contamination. They are able to pack more data onto a given surface area and can hold 5–40 Mbytes of data compared to 0.25–1 Mbyte on a floppy disc. Information transfer rate for floppy discs is typically 10 000–100 000 baud.

Data is stored in short blocks or sectors of the disc. On early disc systems these held 128 bytes. Later systems hold 256 or 512 bytes per sector. The process of formatting marks the positions of sectors and tracks on to the blank disc together with a space for the directory. In simple terms, the detailed way in which data is organized on a floppy disc is set up by the particular computer system. A disc produced on one computer system usually cannot be sensibly read by another computer system without special software. Be warned that formatting a disc destroys all data on it. Do not touch the coated surface of the disc. Do not bend the disc to fit in an envelope. Do not put staples through discs to hold them together. All these things have been tried by the ignorant and do not work!

(iii) *Parallel data transfer*. Transfer of data from the microprocessor takes place on eight wires—the so called data bus. These eight wires carry the data to a variety of peripheral chips. You have already been made aware of memory chips which store and release data when the microprocessor puts the appropriate address on the 16 wire address bus and signals various control wires which will not be further described here. Consider some other chips designed for microprocessors. The UART receives data in parallel from the microprocessor then clocks it out at a rate controlled by the baud rate clock in serial form or else receives serial data and converts it into 8-bit parallel data which is then sent to the microprocessor. The floppy disc controller chip takes

Introduction

parallel data and converts it into a form suitable for writing on the disc. It drives stepper motors controlling the disc speed and the radial position of the head so that a sector address can be translated into an actual head position. By activating a solenoid, the head can be raised or set close to the disc for recording or playback. The parallel input/output device or PIO sets up two 8-bit ports which can act as inputs or outputs and can be used with a variety of further devices. A PIO is often wired to implement a Centronics standard interface. Pins 2−9 carry the eight data bits, pin 1 carries a strobe signal and pin 10 an acknowledge signal. Pin 14 carries the ground. Counter/timer chips can be used to count pulses or to record time as the number of timing waveform pulses. Analogue-to-digital converters (ADCs) and digital-to-analogue converters (DACs) interface with the microprocessor data bus directly or else via PIO chips. Many computer peripherals such as printers and plotters are really small computers in their own right, with their own programs in ROM, and with their own memory store. They receive numbers from an outside device (the microcomputer) and deal with them in some appropriate way. This involves interpreting some numbers as commands to, for example, move the print head to the left. Other numbers are interpreted as printable characters. Either a stamp for that character is selected, for example, by driving a stepper motor to rotate a daisy wheel and the daisy wheel struck against an inked ribbon onto paper, or in the case of a dot matrix printer, certain needles in a pattern of up to 24 (in the most modern printers) are pushed forward by solenoid and again caused to make a mark via inked ribbon. The pattern of dots produces a character. Let us consider the use of byte data for controlling devices such as printers or for representing characters.

3.2 Data, programs and ASCII codes

The 8-bit number is the basic unit stored or manipulated by a microprocessor. Eight-bit numbers are both the data and the instructions.

3.2.1 *ASCII codes*

Few will not have the concept that data input from keyboard or transferred to devices such as terminals and printers appears to be in familiar alphanumeric form rather than as a string of numbers. In reality, the computer 'sees' only the numbers, but external devices are set up to translate the pressing of keys which represent 'normal' characters into 8-bit numbers. In order for the operator to see the characters represented by the key presses in character form the 8-bit numbers have to be translated into appropriate characters seen on the screen or printed by the printer. ASCII (American Standard Codes for Information Interchange) codes are the mappings of 8-bit numbers onto particular characters. Thus A is 65, B is 66, Z is 90. *Figure* 2 shows a common ASCII set. There are small differences between sets, so that 35 which is # in the American ASCII set is £ in the UK set. Not all codes represent characters. Numbers 0 to 31 are usually made control characters of some sort. Thus decimal 13 is generated by the enter key. ASCII 27 (escape) followed by 64 (@) will initialize the default settings of an Epson printer. Numbers from 32 (space) to 127 are familiar alphanumeric characters or punctuation symbols. Codes 128 to 255 are less standard codes representing various characters such as Greek symbols or are sometimes used for limited graphics characters, or for screen control.

	0	1	2	3	4	5	6	7	8	9	A	B	C	D	E	F
0		1	2	3	4	5	6	7	8	9	10	11	12	13	14	15
	0															
1	16	17	18	19	20	21	22	23	24	25	26	27	28	29	30	31
2		!	"	#	$	%	&	'	(	)	*	+	,	-	.	/
	32	33	34	35	36	37	38	39	40	41	42	43	44	45	46	47
3	0	1	2	3	4	5	6	7	8	9	:	;	<	=	>	?
	48	49	50	51	52	53	54	55	56	57	58	59	60	61	62	63
4	@	A	B	C	D	E	F	G	H	I	J	K	L	M	N	O
	64	65	66	67	68	69	70	71	72	73	74	75	76	77	78	79
5	P	Q	R	S	T	U	V	W	X	Y	Z	[	\	]	^	_
	80	81	82	83	84	85	86	87	88	89	90	91	92	93	94	95
6	£	a	b	c	d	e	f	g	h	i	j	k	l	m	n	o
	96	97	98	99	100	101	102	103	104	105	106	107	108	109	110	111
7	p	q	r	s	t	u	v	w	x	y	z	{			}	~	
	112	113	114	115	116	117	118	119	120	121	122	123	124	125	126	127

Figure 2. An ASCII character set showing the printed character (top of box) and the byte coding it underneath. The first 32 characters are not printed. The hexadecimal equivalent of each decimal code may be read from the x and y coordinates of each box.

(i) *Data*. Data can hence be carried in the computer in two ways. Consider the data set 12, 23, 34, 45, 56, 67, 78. This could be carried as 7 bytes, or as seven pairs of bytes containing the ASCII codes for the separate digits, that is, as the following 14 bytes:

49, 50, 50, 51, 51, 52, 52, 53, 53, 54, 54, 55, 55, 56.

In the first case, if you tried to see the data directly on the screen or on a printer, the first 2 bytes would be non-printable and the rest would show as

''−8CN.

In the second case, the data file would show as

12233445566778.

Usually data files are constructed as in the second case as strings of ASCII characters, but with some separator such as carriage return or space or comma between groups of digits constituting a data value.

12,23,34,45,56,67,78

The first method can of course be extended to numbers greater than 255 by using two or more bytes to store a number. A common format for integer data is to use 2 bytes. Integers between 0 and 65535 can be stored in this way.

Often negative numbers are required. It is possible to use 2 bytes of data to represent numbers from −32768 to +32767. Positive numbers are coded in the normal way, using the first 15 bits (0−32767). Numbers above 32767, that is those with the 16th bit a 1, represent negative numbers according to the rule that if $n > 32767$ then $n = n − 65536$.

Introduction

Table 1. Examples of decimal, hexadecimal and binary numbers.

Decimal	Hexadecimal	Binary
123	7B	01111011
45	2D	00101101
255	FF	11111111
110	6E	01101110
172	AC	10101100
48813	BEAD	1011111010101101
250	FA	11111010
50	32	00110010
65530	FFFA	1111111111111010
3200	0CB0	0000110010000000
12345	3039	0011000000111001

This so called 2's complement method is more logical than it might first appear. Consider the numbers $+1$ and -1. Their sum is zero. Let us try this using the 8-bit representations. -1 is represented by 255 or 11111111. The sum of $+1$ and -1 in binary is 00000001 + 1111111 which is the 9-bit number 100000000. In an 8-bit system the 9th bit would overflow and be lost leaving us with zero. Most BASICs on microcomputers which have instructions for accessing 16-bit integers do so using this 2's complement convention. Take care you know how a particular language is dealing with integers at any time.

In some formats and operating systems, a file will have a header block telling the length of the file and other information such as the date it was last accessed. Different microcomputers store 'string' data in different ways.

3.2.2 Hexadecimal: a summary of binary

The novice will quickly encounter hexadecimal numbers. These are numbers to the base 16. In the decimal system it is convenient to use powers of the base such as hundreds or thousands to summarize large decimal numbers. Similarly it is convenient to use hexadecimal numbers to summarize binary numbers. A single hexadecimal number has values from 0 to 15 decimal. This is the same range as a 4-bit binary number. Eight-bit binary numbers can be summarized as two digit hexadecimal numbers.

The first problem is the English language which only has 10 digits. The hexadecimal system uses the numbers $0-9$ and then the letters $A-F$. Hence A Hex is 10 decimal. F Hex is 15 decimal. Otherwise hexadecimal numbers follow the rules for any number system as described earlier in Section 3.1.1. Thus Hex 7FFF is

$$7 \times 16^3 + 15 \times 16^2 + 15 \times 16 + 15 = 32767 \text{ decimal}$$

Its binary value can be found easily by finding the 4 bits representing each hexadecimal number and placing the blocks of 4 bits in the same order as the Hex digits. It is hence 0111 1111 1111 1111 binary. *Table 1* gives examples of binary, decimal and hexadecimal values.

3.2.3 Video RAM

In the simplest form of video mapping, the screen comprising $16-24$ lines of $48-80$

characters is represented by an allocation of memory. Each byte maps to one character position on one line of the display. Changing 1 byte of the memory contents, for example from 67 to 50, would change the displayed symbol from C to 2 in one particular place on the screen. Blanking the screen is done by writing the number 32 (space) into all video memory locations. In more modern colour displays, there is more direct mapping of bits in video RAM to pixels on the screen. Some systems operate separate red, green and blue memory planes. Others allocate 2 or 4 bits per pixcel allowing colour information to be incorporated. In these there has to be a font generation procedure whereby ASCII characters are assigned screen character images as a series of dots. This allows software control of characters allowing different sizes and styles of characters to be used. In early microcomputer configurations from several years back when RAM memory was expensive, and few systems had more than 64 K or 128 K, allocating 32 K or 48 K for a video display was a sizeable chunk out of the useful memory. Machines tended to come with a limited memory mapped display, and high resolution graphics was added as extension boards containing their own RAM and often having a switch mechanism so that the memory used was separate from directly addressed RAM. (See Chapters 2 and 8 for a discussion of IBM graphics boards. Chapter 4 discusses image processing requirements.) More modern machines with ample RAM capacity can easily maintain video RAM in their normal memory space, although many retain separate video RAM which is accessed through a port like other peripheral devices.

3.2.4 *Parity*

The fact that most common alphanumeric characters have codes less than 127 means that really a 7-bit code is sufficient. Most ASCII sets you will see defined in printer manuals, for example, are from 0 to 127. The last, most significant bit is often used in a form of error detection. It is often used as the parity bit in transfer of data to peripheral devices. The parity bit is fixed by the transmitting device so that the number of bits at logic 1 in the byte is odd (odd parity) or even (even parity). The most likely error in transmission is that a single bit will be transmitted wrongly. This will mean that the sum of the bits will go from odd to even or *vice versa*. The receiving device will hence be able to tell that an error has occurred and can demand retransmission. The problem with this sort of system is that it is impossible to directly transmit data and programs which use the full 8-bit range.

3.2.5 *Handshake*

Serial and parallel data transfer methods make use of handshake. Consider a printer. It will receive control codes which set it ready to print in a particular way, then it will receive a stream of printable (if not readable) characters such as this chapter. The data will likely be transmitted from a serial port at 9600 baud, that is approximately 960 characters per second. A parallel Centronics lead will work even faster. The printer will have a buffer of 1−8 K, and this will be able to receive the characters at the transmission rate. Even the fastest of printers do well to deal with 200 characters per second. High quality print is often slow, 30−50 characters per second. Clearly the printer will not be able to deal with the flow of data from the computer. The parallel lead uses a strobe signal from the computer to signify data available. Once the printer

Introduction

has successfully read the data into its buffer, it sends an acknowledge signal to the computer which responds by sending the next byte of data. When the printer buffer is full, the acknowledge signal is not sent until more space is made available, as will happen when the buffered data is printed.

Serial leads provide a selection of handshake signals. Unfortunately no computer manufacturer seems sure of which selection of handshake lines to use, and on which pins of the 25 pin D Connector the lines should be carried. Important pins on an RS232 Connector are 2 and 3, which carry the transmit and receive data respectively, 7 which is signal ground and 1 which is chassis ground. Pins 4 [Request to Send (RTS)], and 20 [Data Terminal Ready (DTR)] are often used to signal buffer full by going from a high level to a low level. Pins 5 [Clear to Send (CTS)] and 6 [Data Set Ready (DSR)] signal data is available. If you are not yet confused, then you probably will be when I tell you that your microcomputer can be wired as Data Terminal Equipment (DTE) in which case it sends data out along pin 2 (TxData) and receives it on pin 3 which is RxData or it can be set up as Data Communication Equipment (DCE). If it is DCE, it sends data out through pin 3 which is still called RxData and receives it on pin 2 which is still called TxData. It is a sign of the confusion that special testing equipment is available which plugs into the serial line and indicates via light emitting diodes (LEDs) which lines are active. It will also allow you to try connecting various handshake lines together until you come up with a working arrangement. The rule with RS232 equipment is that although it may seem easy, if you haven't got a working model to copy, summon an expert with an RS232 tester.

(i) *File control characters.* Handshaking may be done via software, using control characters. The XON/XOFF system sends an ASCII 19 character to stop transmission. Often transmission is restarted on receipt of any other character, although strictly speaking an ASCII 17 character is used to start transmission. Often ASCII files are terminated by a 'Control Z' character, ASCII 26. The receiving device does not need to be informed of the length of the file. Clearly these characters could not be sent as part of the data stream. Such protocol is not suitable for sending the full range of 8-bit characters.

With diminishing costs of memory, it is possible to construct large printer buffers. Software is available for most machines with greater than 500 Kbyte RAM to use several hundred Kbytes as a printer buffer. As an indication of what this means, this book has less than 1 Mbyte characters. Several chapters would fit into printer buffers currently available. Once full, a 256 Kbyte buffer to a printer which prints at 50 characters per second, would take $256 \times 1024/50$ sec or about an hour and three-quarters to empty.

3.3 Machine code programming

The microprocessor can shuttle bytes to and from memory and peripheral devices as indicated above. It can also do much more than that. It contains its own special memory units which are known as registers and can hold 8-, 16- or 32-bit numbers depending on the particular microprocessor chip. Numbers can be manipulated arithmetically, have bitwise operations performed on them and can be stored under unique addresses or in temporary stacks. The microprocessor reads an instruction which may be one or several bytes long from memory, and then executes that instruction. It then takes the next instruction from memory and executes this and so on. A program consists of sequential sets of instruction bytes in memory. Instructions may involve jumps to particular memory addresses (absolute addressing) or jumps a certain number of bytes ahead or behind the current position in the memory (relative addressing). Any program which contains jumps to specified memory locations must always reside in the same place in memory. Programs with only relative addressing can reside anywhere in memory and are said to be *relocatable*. Fetching instructions and executing them is done under the strict time control of a square wave clock, with each clock cycle setting in motion part of the instruction sequence. Typically the clock runs at 2 – 12 MHz (million cycles per second).

3.3.1 *An introductory machine code program*

The numbers representing instructions do not mean much to us. Consider the hexadecimal numbers

21 00 10 DB 04 77 23 DB 04 77 23 DB 04 77 23.

You cannot even guess how many different instructions there are here, although you can begin to see a repeating sequence. To simplify things each instruction is given a so called *mnemonic* which is an abbreviated, stylized description which yields understanding of function. Consider the following mnemonics and hexadecimal codes for a Z80 microprocessor:

LD HL,nn	21 nn nn
IN A,(n)	DB n
LD (HL),A	77
INC HL	23

With specific numbers and some explanation:

LD HL, 1000H	21 00 10	Load the 16-bit register called HL with the number 1000H. Note the code contains the number low byte first.
IN A, (04)	DB 04	Input an 8-bit number from port number 4 (which could, for example be an eight wire data set from an ADC) into the A register.
LD (HL),A	77	More abstract this because the end result in this case is to load memory location 1000 H(exadecimal) with the value in the A register. The instruction means load the address pointed to by the contents of the HL register with the contents of the A register.
INC HL	23	Increment the contents of the HL register. This will now contain 1001H.

3.3.2 *Time per instruction*

Each time the sequence DB 04 77 23 is repeated, another byte is read from port 4 and stored in memory sequentially from address 1000H upwards. How long would these instructions take to execute on a 4 MHz Z80 based microcomputer? The manufacturers provide data in terms of the number of time states or clock cycles as shown in *Table 2*.

Introduction

Table 2. Times per instruction.

Mnemonic Hex	Code	Time states	Time (μsec) (4 MHz Z80A)
LD HL,nn	21 nn nn	10	2.5
IN A, (n)	DB n	11	2.75
LD (HL),A	77	7	1.75
INC HL	23	6	1.5

Table 3. An alternative program with instruction times.

Mnemonic	Code	T states	Time (μsec)	Explanation
LD C,04	0E 04	7	1.75	Loads the value 4 into the C register
LD HL,1000H	21 00 10	10	2.5	HL register set to 1000H
INI	ED A2	16	4	Loads the address pointed to by the HL register with the contents of the port pointed to by the C register, i.e. 4 in this case. The B register is decremented to count the number of times the operation is performed.

It takes 6 μsec (2.75 + 1.75 + 1.5) to take a byte from the port and store it in memory. An ADC supplying its data to port A would be read every 6 μsec or 166 666 times per sec. A simple working program could be as follows:

21 00 10 DB 04 77 23 DB 04 77 23 . . . . DB 04 77 23

where the dots represent the sequence DB 04 77 23 repeated another 997 times representing 1000 bytes in all read from the port. The program would be unnecessarily long (4003 bytes), because a loop with loop counter could reduce the length of the program which is clearly highly repetitive. It would however be code optimized for time for the given limited sequence of codes. In fact there is a single instruction which works slightly differently which could have had the same effect.

3.3.3 Another program with the same effect

Consider the codes in *Table 3*. The program 0E 04 21 00 10 ED A2 ED A2 . . ED A2 where the dots represent the instruction ED A2 repeated 997 times carries out the same function as the one before, this time with a loop time of 4 μsec, or a possible sampling rate of 250 000 Hz. This would be a time optimized machine code program to input 1000 bytes of data from a port to memory. The code in this case would be 2005 bytes long. Two very different machine code programs hence achieve the same sort of effect. The program utilizes 1000 bytes of memory from 1000H upwards, that is from 1000H to 13E8H for data storage. Otherwise the program itself could reside anywhere in the 64 K RAM available to the Z80 microprocessor. It is relocatable code.

3.3.4 *Entering and using machine code*

Machine code programs such as the one above could be written into memory in a microcomputer using a special monitor program which allows memory contents to be displayed and altered, such as DDT supplied with CP/M operating system computers or NAS-SYS3 on Nascom microcomputers, or less directly using higher level language instructions such as PEEK or POKE in BASIC which allow the user to see and change a byte at a particular address. In general, early microcomputers which preceded higher level languages such as BASIC, came with a monitor program which allowed convenient entry, alteration and execution of machine code programs. Later machines steer the user clear of machine code by not providing a simple method of altering memory contents. More conveniently, machine code programs are constructed using a special program called an *assembler* which accepts a file of mnemonics and then looks up the appropriate code for each, and constructs a sequential file of the numbers. Special assembler directives or instructions allow the user to set up the start address for the program. (See *Listings 1* and *2* in the Appendix to this chapter.) Most assemblers also allow the user to combine several small programs into a bigger one. In general using machine code is a specialized skill which is now unnecessary for all but certain interface tasks where strict control of timing is required. Nevertheless, a basic knowledge is extremely useful. Often it is a trivial machine code programming task to change a few bytes in a program to allow extended use of the program. This can often be done from little knowledge of the program.

3.3.5 *Interrupts*

Microprocessors have special input lines which when switched active cause the microprocessor to stop (interrupt) its steady execution of sequential instructions, temporarily store the address of the next instruction which would have been executed had the interrupt not occurred and jump to a new memory location (the start of a different program). This new program is then executed until a special return from interrupt instruction is encountered. The microprocessor then recovers the address of the instruction temporarily stored when first interrupted and proceeds with the original program.

Consider the simple example program given in Section 3.3.1 above. It takes in a byte which could be from an ADC every 6 μsec. For physiological use this is excessively fast; most physiological data logging of the sort previously done by pen recorder requires conversion times of from 1 to 1000 msec. Even sampling a rapid event such as an action potential only requires sampling every $50-100$ μsec. It would be possible to pad out our program using neutral instructions such as NOP which does nothing for 1 μsec. This would be inconvenient to program and wasteful of program space and processing time. Note it is possible to use the instruction execution times as the program timer (see *Figure 3*) and this may be a way to proceed when resources are limited. A more convenient way of proceeding would be to arrange to activate the interrupt line to the microprocessor chip every 1 msec or whatever time was required as sampling interval.

Introduction

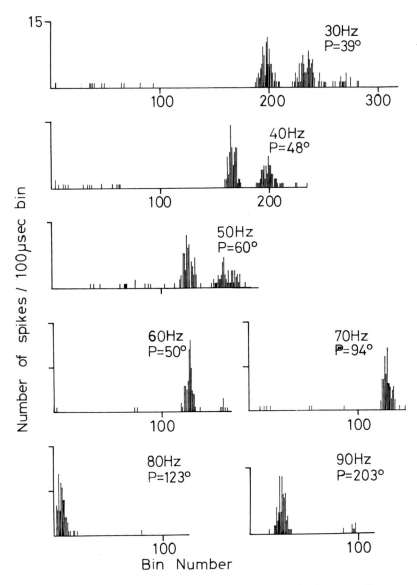

Figure 3. Histograms of numbers of spikes per 100 μsec bin for crab balancing organ afferent neurons at different oscillation frequencies. These were obtained using a simple program which used a loop of Z80 instructions padded out to last exactly 100 μsec. During the loop, the state of 2 bits of a port were monitored, and a bin number pointer incremented. The state of 1 bit indicates the beginning of an oscillation cycle and causes a jump to a routine which resets histogram pointers and tests for the end of the required number of cycles. A pulse to the other, from a spike trigger, causes a jump to a routine where histogram bin contents are incremented. By careful monitoring of the time taken for each instruction, all timing information can be derived from the microcomputer itself.

The interrupt program could then consist of the instructions to take in a byte, store it in memory, increment the memory pointer and return from interrupt. Special purpose chips like the Z80 family counter/timer circuit (CTC) (e.g. Mostek MK3882 CTC) can be programmed to down-count the system clock (4 MHz) pulses from a preset level until the count reaches zero when an interrupt pulse is produced and sent to the Z80 microprocessor.

In the following section, some programs used in studies on equilibrium and auditory systems are presented to give an idea of the steps involved in interfacing a microcomputer in the physiological laboratory.

4. LOW BUDGET APPLICATION OF MICROCOMPUTERS TO EQUILIBRIUM RESEARCH

4.1 Interface requirements

The balancing organ of the crab is analogous in many respects to the vertebrate semi-circular canal system, and may be investigated in the frequency domain (6). In order to investigate the neurophysiological basis of equilibrium control in the crab, it was necessary to be able to oscillate the crab from 0.1 to 100 Hz at a range of amplitudes. Recordings from nerve spikes and electromyograms (EMGs) from muscles involved in equilibrium reactions had to be processed and analysed with reference to the imposed angular accelerations. At the start of this project, the Nascom 3 microcomputer (Lucas Logic) was the most suitable choice of machine. This had 48 Kbytes RAM, an 8 K ROM BASIC and used tape recorders to store programs and data. Dual floppy discs were added later. An input/output (I/O) board housed 3 MK 3881 parallel I/O controllers and 1 MK 3882 CTC. With the onboard PIO, this gave four PIOs and one CTC for interface purposes. Machine code could be entered directly using NAS-SYS3 Monitor commands, or via a Z80 assembler (PolyZap) with files prepared on the editor PolyEdit supplied as part of the PolyDos disc operating system.

Although the Nascom is an early generation microcomputer it has proved an excellent and reliable interface machine. Data obtained on Nascom is now transferred via a parallel link to a Sinclair QL expanded up to 896 Kbytes RAM, with twin 3.5 inch floppy discs, each holding 720 Kbytes formatted data. Data is further processed using compiled BASIC programs (Supercharge, Digital Precision), allowing graphics display and print out.

4.2 Servomotors and D/A converters

Inland DC servomotors were chosen to oscillate the crab. A T-2967-A servomotor and TG-2917-D tachometer generator were driven by a 200 W EM-1800-00-A servo amplifier (Inland Motor) (see ref. 7). In the adopted configuration, a signal of $+/- 5$ V input to the servoamplifier, was sufficient to give the range of oscillation amplitudes suitable for horizontal and vertical axis angular accelerations in the above range. To drive the motor, a 16-bit latched DAC with a 14-bit DAC (Hybrid Systems) providing the reference voltage was used, so that the computer could control the driving voltage, and the gain of the voltage over a wide range using four 8-bit ports from two PIOs (*Figure 4*). Position around the vertical axis can be controlled using a DC servomotor system and controller (RS Components, 336-309, 591-663 and 336-264).

4.3 Timing and spike counting with CTC

The counter/timer circuit is a four channel device. Each channel occupies a single port, and consists of an 8-bit counter (down-counter) which can be read by the host computer

Introduction

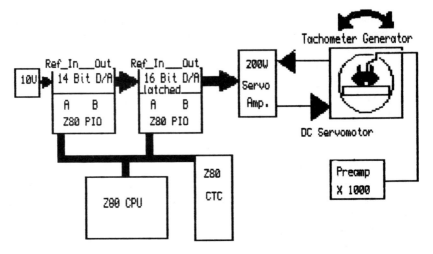

Figure 4. The computer controlled oscillator uses a 16-bit latched DAC to output a 100 point sine wave stored in the microcomputer memory. This output sine wave is used to oscillate a crab via a servomotor system. A CTC sets the time for each part of the sine wave to be output. A 14-bit D/A is used as the reference voltage to the 16-bit D/A, allowing simple control of gain.

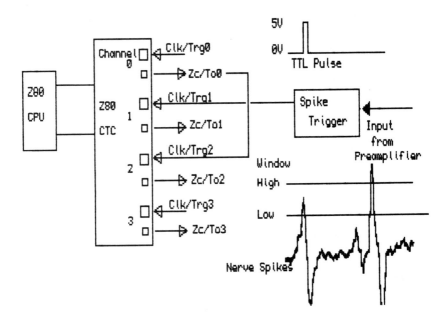

Figure 5. Counter/timer circuit for counting spikes and general timing. Nerve spikes are conveniently hardware processed using a spike trigger such as the Neurolog NL 200 (Digitimer Ltd). This allows high and low thresholds to be set. TTL pulses may be obtained for units which cross the upper or lower thresholds, or as given here, from those spikes which cross the lower threshold, but do not cross the upper threshold within a certain period of time. In the program given in *Listing 1* (see Appendix), the nerve spikes go to channel 1 of a Z80 CTC. Channels 0 and 2 are linked to provide timing information.

using a port input instruction. Control instructions to the CTC are written to the port. Each channel can be software configured as a counter or a timer. In timer mode, the system clock is divided by 16 or 256 and then caused to down-count the register. Each channel can be caused to count down from a preprogrammed value by outputting a time constant to the port. Each channel can be set to cause an interrupt when it counts or times down to zero. In addition, on reaching zero, a pulse is output from the zero count/time out pin. This can be used to clock a second channel.

4.3.1 Using the CTC in the oscillator program

Figure 5 shows the timing arrangement in the oscillator program (see *Listing 1*). Channel 0 is set up as a timer, to divide the system clock by 16, and with a time constant of 25. It hence counts to zero every 100 μsec. The zero pulse output from here is used to clock channel 2 which is set to counter mode and to cause an interrupt when it reaches zero. Different times, from 100 μsec to 256 × 100 μsec can be selected by changing the time constant to this channel. In the program, there is an interrupt for each bin of the histogram, and, concurrently, each of the hundred digitized sine wave values, so a single cycle can be set to last from 10 to 256 msec, corresponding to oscillation frequencies of 3.9−100 Hz. By changing the time constant of channel 0 to its maximum of 256 and programming it to divide the system clock by 256 instead of 16, frequencies down to approximately 0.0024 Hz can be generated. Clearly lower frequencies could be obtained by using the spare CTC channel or more than 100 bins in the histogram. To count spikes, pulses are generated by a spike trigger which sets an upper and lower discrimination window and outputs pulses when a spike crosses the lower window, but fails to cross the upper window within a certain time. These are simply output to the appropriate pin of the CTC.

4.4 A pointer to machine code required

The steps involved in using the computer to output a given signal such as a sine wave to the DC servomotor are as follows.

(i) Using BASIC, generate a 100 point sine wave using the range 0−65535, that is oscillating between +32767 and −32767.
(ii) Store this as pairs of bytes in memory. These clocked through the D/A in sequence will generate a good sine wave. Further filtering to get rid of the small steps in the trace, gives a cleaner looking output signal, but makes little difference to the output of the motor, which acts as a low pass filter anyway.
(iii) Using the editor and assembler, write a routine (see *Listing 1* in Appendix) which will satisfy the requirements of *Figure 6*, and output the stored sine wave values. *This is not a trivial task.*
(iv) Connect up DACs to the ports, and test their operation using an oscilloscope to monitor the output.
(v) Test spike pulses by using a stimulator to mimic pulse trains between 1 and 1000 Hz.
(vi) Use a series of clocking times to test operation over the frequency range.

Introduction

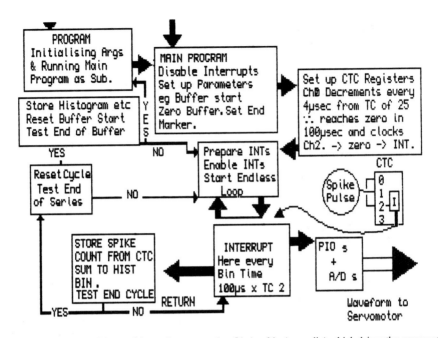

Figure 6. Main parts of the machine code program (see *Listing 1* in Appendix) which drives the servomotor and averages the nerve spikes. One channel of a CTC is set to count to zero every 100 μsec. The zero count pulse from this clocks a second channel, which down-counts from a programmed initial value (time constant). When this reaches zero, an interrupt is generated to a routine which outputs the digitized sine wave to the servomotor via a PIO and DAC. Nerve spikes, counted in one channel of the CTC are summed into the appropriate bin of the histogram. At the end of a cycle, the program tests for the end of a number of cycles, and resets bin pointers. At the end of a series, the histogram and associated information is stored.

(vii) Arrange for a higher level language such as BASIC or Pascal to alter parameters such as gain and frequency. Further test operation.

4.4.1 *Software subroutines*

The task of writing the software (step iii) is best split into small easily tested parts. It is convenient (if you cannot simply copy a working routine such as is given here) to test the CTC by using a simplified interrupt routine which increments a location in video RAM and returns from interrupt. Once the routine works, the other subroutines can be added. These again are written and tested in isolation. Even so, the time involved in testing the final program is considerable. It is an extremely confident programmer who would claim error free software. Errors can be extremely subtle. If, as is likely, you will not be writing machine code software yourself, but relying on professionally produced software, be aware of the possibilities of errors. Test your programs rigorously and make suggestions for upgrading to the author. Go back to your software supplier after a year, and ask if the software has been upgraded.

4.4.2 *Oscillator program*

Listing 1 of the oscillator program (see Appendix) has been annotated to try to make

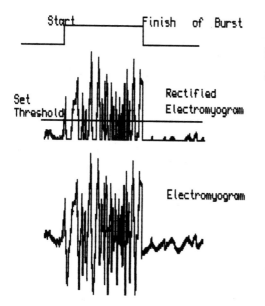

Figure 7. Recordings from muscles are rectified and digitized at 1 msec intervals. A threshold is set in software, and a number exceeding threshold is taken as the start of a burst and the time noted. Successive digitized values are summed to provide the integrated activity. When 100 successive samples are below threshold, the burst is considered to be terminated, and the time of the last reading above threshold is taken as the finish time. Start time, finish time and integrated activity are stored for each burst.

it understandable to an inexperienced programmer. It is not possible here to go into detailed explanation of how to program PIOs and CTCs. Clear accounts are given in the manuals for the chips which are available from manufacturers such as Mostek. If you are keen to start on machine code, find out about assemblers for your particular machine and ask for introductory textbooks for the particular language involved.

4.5 Using an ADC to monitor muscle potentials

A more complicated example of machine code programming is given in *Listing 2*. This muscle potential analysis program listing also shows the code generated on assembling the program. It is a fairly large program which was derived from a more general routine for timing two channels of pulses (nerve spikes and events). The DC servomotor system described in the previous sections is used to simply output a positive pulse, followed by a negative pulse to generate the angular acceleration required to displace a tethered crab in the head-up direction followed by displacement in the head-down direction. It is hence possible to automate the displacement protocol used to study tilt evoked swimming in crabs (8). The computer is further used to rotate the crab around the vertical axis and digitize two EMG channels and store start times of the bursts, finish times of the bursts and the integrated activity (*Figure 7*). The program is used in ongoing work on long term monitoring of tilt evoked responses. It has been used to store EMG parameters for up to a fortnight at a time to tilts every 15 min in the planes of left and right statocyst vertical canals. *Listing 3* is the Pascal source program which is used to run the program. The compiled Pascal code is combined with the run

Introduction

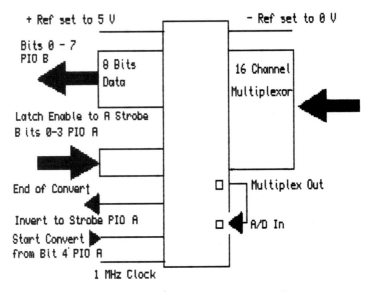

Figure 8. Main connections of 16-channel multiplexing ADC (ADC 0816).

time package and the code for the muscle analysis program to give a single machine code program.

A working rule from vestibular and auditory laboratories is that 90% of data processing should be done on line. Consider a muscle burst lasting 100 msec. This could be stored as 100 samples (digitized every millisecond). Using the analysis program, we lose information, but retain start time, finish time and integrated intensity as 7 bytes of data (3 bytes start time, 2 bytes finish time and 2 bytes intensity).

It is not expected that the beginner should follow everything that is happening in the program, and even a simplified flow diagram would be too complicated. The full listing is given for experienced programmers and to give a proper indication of the complexity of machine code programming. The program has evolved from a pulse timing program and has been deliberately structured to be easily altered by the author for a wide range of stimulus and recording situations.

4.5.1 Using a multiplexing ADC

A multiplexor is a software controlled switch which allows several input lines to access a single ADC, for example in rapid succession. In the example given in *Figure 8*, 16 input lines carry data to the multiplexor. These are switched to the multiplex output line under the control of a 4-bit address register which is supplied by port A of a PIO. Writing out the 4-bit word 15, for example, will switch channel 15 to the output. This 4-bit address data is latched into the PIO with the PIO A strobe.

The multiplexor may be used separately from the ADC, but is usually wired to the A/D input pin. Start of conversion is controlled by sending a pulse to the start convert pin. Any control pulses can be software generated using spare PIO bits so that here, setting bit 4 of PIO A high and then low, starts conversion under convenient programming control. The end of convert signal is connected to the PIO B strobe line, so that the 8-bit data output from the ADC is latched into port B of the PIO. (The strobe line acts as a sort of switch to catch and fix the data into a special memory register in the PIO. It can then be read at any time.) A 1 MHz clock signal is required to clock conversion at its fastest rate to give a conversion time of 100 μsec.

4.5.2 Outline of muscle analysis program

The program displaces the crab head-up for several seconds, then displaces it head-down. This is repeated five times, the crab is then rested for 7.5 min and rotated using a DC servomotor controlled by an 8-bit DAC. Succesive rotations position the crab in the planes of left and right statocyst vertical canals, respectively (which in turn mean that the statocyst information is limited to known single cells running from the brain to the lower ganglia). Electromyograms are recorded from two leg muscles, filtered (band pass 20–200 Hz), rectified and fed into channels 0 and 1 of a 16-channel multiplexing 8-bit ADC. The program sets a threshold to operate on the digitized data as it is received every millisecond. When a value exceeds threshold, an 'in-burst' pointer is set, and the time noted. Successive digitized values are summed into an 'integrate' memory location. When 100 msec have elapsed with no values above threshold, the burst is considered terminated and the time of the last value above threshold noted. Start time, finish time and integrated activity are hence recorded.

4.5.3. Use of CTC in muscle analysis program

Timing information is supplied simply using a single channel of the CTC, set to time and interrupt every millisecond. A/D channel 0 is selected and conversion started. After a wait of 200 μsec, a 3-byte time counter is used to initiate pulses through the 16-bit DAC which are translated into angular accelerations sufficient to cause a displacement of the crab. The last time signifies the end of one cycle. The time register is reset, and marker bytes inserted into the data.

We have now stored the A/D values for both channels. For each channel, if an in-burst marker is set, the A/D value is summed into a 2 byte intensity register. If above threshold, the time is stored as a record of the last time an above threshold value was read. If not above threshold, the current time is compared with that record, and if 100 msec have elapsed, the in-burst marker is reset to zero and the last time above threshold gives the end of the burst. If the in-burst marker is not set, then the A/D value is compared with the threshold. If above, the time is stored as the start of a burst and the in-burst marker set. The program was written for one channel and the assembler program largely repeated for the second channel, but postfixing variable names with '1'.

The pulse routine used in modified versions of the program has been disconnected by adding a RETI (return from interrupt) instruction immediately. Spikes are converted to TTL pulses and sent to channel 2 of the CTC which is set to count down from 1 and interrupt. Each spike hence causes an interrupt to a routine which accumulates a spike count in 1 byte in an allocated memory block. This feature is not used in long term muscle recording, so the buffer capacity of this has been reduced to just over 1 min. The program is intended for modification using higher level language for simultaneous nerve spike and EMG recording, for a reduced number of cycles.

Introduction

4.6 Using a higher level language to control a machine code module

Higher level languages are easy to program and convenient to use. However they cannot be used for applications where strict timing is necessary. A good compromise is to use a mixture of higher level language and machine code for interfacing, data logging and data reduction tasks, and to use the higher level language for subsequent data processing. In BASIC, the instructions PEEK and POKE allow access to single bytes at specified addresses. Consider *Listing 2*, near the beginning there are instructions:

LD A, 095H
OUT (8),A

When assembled they give the following (memory location followed by bytes):

2B2B 3E 95
2B2D D3 08

The byte at memory location 2B2C Hex is 95 Hex. It is a control instruction to channel 0 of the CTC configured to ports 8−0AH. It sets it to interrupt and divide the system clock by 16, that is to use 4 μsec timing intervals. If that byte were changed to A5 Hex, then it would now be the control word meaning set to interrupt and divide the system clock by 256 (see the following instruction in the listing). Our whole program would now use 64 μsec timing intervals instead of 4 μsec intervals. The ADCs would now sample at 16 msec intervals. The BASIC instruction

nnn POKE 11052,165

could hence change our sampling rate by a factor of 16. Similarly, changing the time constant value at address 2B4A from 250 to something else would change the timing.

It is convenient to use BASIC instructions also to call the machine code program as a subroutine terminated by a RET (C9H) instruction. Timing parameters, loop counters and buffer start or finish are all conveniently manipulated using POKE instructions, or inspected using PEEK instructions. In a similar way, Pascal can be used for control and manipulation. *Listing 3* is an example of a Pascal program which is used to manipulate the machine code routines given in *Listing 2*. The program is used to swing crabs set to different angles around the vertical axis (hence the obscure name SWCBAN10). The '10' gives an indication of the proliferation of versions which appear as errors are corrected, or more and more ambitious steps are added. By using the same variable names for addresses defined in the assembler program, communication between the Pascal and machine code is simplified. Machine code routines which are set up as subroutines are called by defining a procedure name followed by the directive EXTERNAL and the address or constant defined as the address of the routine. For example:

PROCEDURE OSCILLATE;
EXTERNAL ORGIN;

where ORGIN, defined as a constant equal to 2B00H, will call and run the main routine given in *Listing 2*. Other routines call subroutines which allow delays of multiples of a second to be specified, or allow setting of the threshold levels for each ADC. Note that this is best done using real time feedback. In the assembler subroutines ADTRIG0 and ADTRIG1 the threshold level can be incremented or decremented using up arrow and down arrow keys.

The procedure SETANGLE is an example of a higher level language program used directly to control an external device. It simply accepts angles of -45, 0 and 45 degrees and outputs the appropriate numbers (127, 165 and 202) which when fed out of port 1DH to an 8-bit DAC, generate an appropriate control voltage to make a DC servomotor turn the crab around the vertical axis to the appropriate angle. Note this function could easily have been made more general by incorporating a function calibrating the angle produced for a given value at port 1DH. A delay of 30 sec allows the crab to be turned and to settle down. Procedures STORE, WRIT, CALCDATASPACE AND NOTROOM check for buffers full and if full, pass data via a serial link to a disc based computer system. Other procedures allow a single cycle of oscillation at $+45$ and -45 degrees, or arrange to store data after 12 cycles (3 h). With normal activity, approximately 160 Kbytes of packed data are collected every day. The program displays a menu allowing selection of a single oscillation series, continuous oscillation, or resetting of threshold levels for the two muscle channels. This requires real time feedback which is provided in machine code, by the display of a solid block character when threshold has been exceeded (i.e. in-burst) or a space when out of the burst.

The compiled Pascal program can be combined with its runtime package (the minimal set of machine code necessary to run a Pascal program) and the machine code routines to create a compact machine code program. The compiled program is about 20 times faster than an equivalent BASIC program for the Nascom. Combined machine code and higher level language programs are powerful units for experimental control, data acquisition and data transfer. This chapter has not addressed the problem of what to do with the data once obtained via interfaces, and this is not a trivial task as can be seen in the next chapter.

4.7 Good and bad programs

A good program is one that works and which does not stop any other program from working. Good programming practice with assembler and machine code is to save all registers which are used by the program and restore them before returning. Provided the program has not affected any area of memory used by another program in an improper way, all will be well. It is necessary to inactivate or restore to their previous state, all programmable external circuits such as PIOs and CTCs. I have never used a computer system yet which does this properly. Most of the example programs given in this chapter are bad programs in this respect. The way the programs are used, however, this does not matter. They are reliable within their normal operating constraints. Be careful when you ambitiously string several programs together on a computer. Things may happen which do not occur when either is run on its own from cold. This should not happen, but it does. All programs and supporting hardware must be rigorously tested in its own application configuration. This is good practice for any new program, since the use of your testing regime will give familiarity and improve speed and accuracy.

Introduction

5. CONCLUSION

Although interfacing is now properly the domain of the professional laboratory supplier, some understanding of machine code methods and strategies will help the physiologist use and develop interfaces and their applications software. Much is possible with a little knowledge and a few electronic components. Often, direct linking of port pin to peripheral chip pin is all that is required. In the rest of the book, machine code and interfaces are less emphasized.

6. REFERENCES

1. Mayzner,M.S. and Dolan,T.R.(eds) (1978) *Minicomputers in Sensory and Information-Processing Research*. Lawrence Erlbaum Associates, NJ, USA.
2. Bures,J., Krekule,I. and Brozek,G. (1982) *Practical Guide to Computer Applications in Neurosciences* J.Wiley and Sons, NY, USA.
3. Ireland,C.R. and Long,S.P. (eds) (1984) *Microcomputers in Biology—A Practical Approach*, IRL Press, Oxford, UK.
4. Kerkut,G. (1985) *Microcomputers in the Neurosciences*. Clarendon Press, Oxford, UK.
5. Mize,R.R. (ed.) (1985) *The Microcomputer in Cell and Neurobiology Research*. Elsevier, NY, USA.
6. Fraser,P.J. (1981) In *Vestibular Function and Morphology*. Gualtierotti, T. (ed.) Springer Verlag, NY, USA, p. 206.
7. Tako,S.L. (1979) *Understanding DC Servoamplifiers*. Machine Design, Inland Motor, Kollmorgen Corporation, VA, USA.
8. Fraser,P.J., Bevengut,M. and Clarac,F. (1987) *J. Exp. Biol.*, **130**, 305.

P.J.Fraser

APPENDIX

Program listings

Listing 1. Program to control DC Servomotor used to oscillate crabs, and record nerve spikes as a cycle histogram.

```
;OSCILLATOR PROGRAM

;PROGRAM OSCILLATOR USING 16 BIT LATCHED D/A TO PRODUCE DRIVING
;WAVEFORM. 14 BIT D/A IS REFERENCE VOLTAGE TO THIS TO CONTROL GAIN
;D/AS ARE INTERFACED VIA PIOS AT PORTS 4 AND 5 AND 18H AND 19H
;CTC IS USED FOR TIMING AND FOR COUNTING NERVE SPIKES.
;LINK ZERO COUNT/ TIME OUT 0 TO CLOCK/ TRIGGER 2 OF Z80 CTC.
;TO CONTROL TIMING. TTL PULSES FROM A SPIKE TRIGGER GO TO CLOCK/
;TRIGGER 1 WHICH IS SET TO DOWN COUNT FROM 256 EVERY BIN PERIOD.
;NB CTC USES 4 PORTS 8, 9, 0AH AND 0BH. PIOS USE 2 PORTS EACH FOR DATA
;AND 2 PORTS EACH FOR CONTROL.

;NASCOM SPECIFIC FEATURES
;A CONVENIENT MONITOR OF THE PROGRAM IS TO CHANGE A BYTE IN THE TOP
;LINE OF THE VIDEO RAM. THIS SHOWS AS A CHANGING CHARACTER ON SCREEN.
;LOCATION 0BD0H IS SUITABLE ON THE TOP LINE OF NASCOM VIDEO DISPLAY.
;THE PROGRAM TAKES ARGUMENTS FROM ARGS 2-6
;ie 0C0EH -0C18H  ON NASCOM THESE HAVE TO BE SUPPLIED.
;IT IS ALSO NECESSARY ON NETWORK MICROS WITH
;BROADCASTING DISABLED TO DISABLE PORTS 4 AND 5
;WITH 03 SENT TO THE CONTROL PORTS 6 AND 7

;ARG2    0C0EH    3086     64       NO. CYCLES
;ARG3    0C10H    3088     100      NO. BINS
;ARG4    0C12H    3090     23000    BUFFER START
;ARG5    0C14H    3092     -1       GAIN
;ARG6    0C16H    3094     0        FREQUENCY (TC)

ORGIN    EQU 2C00H

         IDNT ORGIN, ORGIN    ;ASSEMBLER DIRECTIVE DEFINING ORIGIN
                              ;AND EXECUTION ADDRESS.

CTC      EQU 8
VIDEO    EQU 0BD0H            ;BYTES HERE IN VIDEO RAM AS INDICATOR
GAIN     EQU 18H
DTOA     EQU 4
ARG      EQU 0C0EH            ;ARGUMENTS USED WHEN EXECUTING FROM MONITOR

DATFIN   EQU 0C000H           ;TOP OF USEABLE MEMORY IN NASCOM
BUFFIN   EQU DATFIN-230H
SIGSTRT  EQU 3000H
ARG2     EQU ARG
ARG3     EQU ARG+2
ARG4     EQU ARG+4
ARG5     EQU ARG+6
ARG6     EQU ARG+8

         ORG ORGIN            ;PROGRAM TO RUN MAIN
                              ;ROUTINE AS SUBROUTINE. ALSO INITIALIZES ARGS
```

Introduction

```
                LD HL, 16
                LD (ARG2),HL
                LD HL, 100
                LD (ARG3),HL
                LD HL, 3000H
                LD (ARG4),HL
                LD HL, -1
                LD (ARG5),HL
                LD HL,0
                LD (ARG6),HL    ;ARGS INITIALIZED
                CALL SETCAL     ;CALL INITIALIZATION ROUTINE
                CALL ORGIN+30H
                DB 0E7H         ;SPECIAL INTERRUPT ROUTINE USED IN NASCOM
                                ;TO DISPLAY REGISTERS AND RETURN TO MONITOR
                                ;PROGRAM ALLOWING KEYBOARD ENTRY. COMPUTER
                                ;MAY NOW BE USED NORMALLY WHILE INTERRUPT
                                ;DRIVEN OSCILLATOR PROGRAM CONTINUES.

;END OF PROGRAM TO RUN MAIN ROUTINE AS SUBROUTINE

;START OF MAIN ROUTINE

                ORG ORGIN+30H   ;START OF MAIN PROGRAM, CALLED AS SUBROUTINE
                                ;AS IS REQUIRED OF MACHINE CODE ROUTINE
                                ;CALLED FROM HIGHER LEVEL LANGUAGE.

                DI              ;DISABLE INTERRUPTS. NECESSARY BEFORE CHANGING
                                ;THINGS.
                PUSH AF         ;SAVE REGISTERS ON STACK
                PUSH HL
                PUSH DE
                PUSH BC
                PUSH IX
                PUSH IY

                LD A,03         ;FIRST CONTROL INSTRUCTION (03)TO PIOS
                OUT(6),A
                OUT(7),A        ;DISABLES PORTS 4 AND 5

                LD A,01         ;CONTROL BYTE TO DISABLE CTC BEFORE ALTERING IT
                OUT(CTC),A
                OUT(CTC+1),A
                OUT(CTC+2),A
                OUT(CTC+3),A    ;DISABLES CTC

                CALL SETUP      ;INITIALIZES THINGS
                LD HL, CONT1    ;DUMMY RETURN FROM INTERRUPT WHICH ENSURES
                                ;PROPER FUNCTIONING OF INTERRUPTS
                PUSH HL
                RETI
        CONT1   LD A, HIGH VECAD    ;SETTING UP POINTER TO TABLE OF 4
                                    ;ADDRESSES FOR CTC INTERRUPT
                                    ;ROUTINES.
                LD I,A
                IM 2            ;INTERRUPT MODE 2
                LD A, LOW VECAD ;LOW BYTE OF TABLE KEPT BY CTC
                OUT(CTC),A
                LD A, 05H       ;CONTROL BYTE TO CTC - TIME, NO INTERRUPT
                                ;SYSTEM CLOCK (4MHZ) / 16 - SENT TO
                OUT (CTC),A     ;CHANNEL 0 OF CTC
                LD A,0D5H       ; - COUNT, INTERRUPT - SENT TO
                OUT (CTC+2),A   ;CHANNEL 2
```

```
              LD A, 55H          ; - COUNT NO INTERRUPT - SENT TO CHANNEL 3
              OUT (CTC+3),A      ;SPARE FOR THIS PROGRAM.
              LD A,0
              OUT (CTC+3),A
              LD A, 25           ;TC OF 256
              OUT (CTC),A        ;TC FOR 100 MICROSEC
    TC        LD A, 0            ;TC CHANGED AT GETTC
              OUT (CTC+2),A      ;TC SET TO DEFINE BIN TIME
                                 ;PERIOD = BIN NO X 0.1 X (TC+1)
              LD HL,SIGSTRT-1    ;START OF SINEWAVE
    CONTINUE          EI         ;ENABLE INTERRUPTS
    LOOP      JR LOOP            ;PROGRAM GOES INTO ENDLESS LOOP, JUMPING
                                 ;TO ITSELF. IT IS USUALLY PRUDENT TO CONFINE
                                 ;THE PROGRAM DURING CRITICAL TIMING ROUTINES
                                 ;ALTHOUGH IN A PERFECT WORLD THIS WOULD NOT BE
                                 ;NECESSARY.

;INTERRUPT ROUTINE CALLED WHEN CTC CHANNEL 2 HAS COUNTED DOWN TO ZERO
;IE EVERY BIN PERIOD

              CALC    ORG ((CALC/8)*8)+8
              PERBIN  INC (IY+4)     ;VIDEO MOONITOR
                      CALL SIGOUT    ;CALL ROUTINE TO OUTPUT NEXT VALUE OF
                                     ;WAVEFORM TO OSCILLATOR
                      CALL STORESIG  ;CALL ROUTINE TO STORE SIGNAL AS SUMMED BIN
                                     ;CONTENTS
                      DJNZ CONT2     ;GO TO CONT2 AND RETURN FROM INTERRUPT
                      POP DE         ;IF END OF BIN COUNT GO TO SET UP NEW CYCLE
                      LD DE, NEXTCYC
                      PUSH DE
                      DI
                      RETI
              CONT2   EI
                      RETI

;END OF INTERRUPT ROUTINE CALLED EVERY BIN PERIOD

;ROUTINE TO GIVE WAVEFORM TO DRIVE OSCILLATOR

              SIGOUT  LD A,(HL)       ;SINE WAVE HIGH BYTE
                      OUT (DTOA+1),A  ;1ST BYTE TO LATCHED
                                      ;16 BIT DTOA
                      DEC HL
                      LD A,(HL)
                      OUT (DTOA),A    ;2ND BYTE LATCHED OUT
                      DEC HL
                      RET

;END OF ROUTINE GENERATING WAVEFORM VIA D/A

;ROUTINE ONLY ACCESSED AT THE END OF A SINGLE CYCLE.

              NEXTCYC LD A,(VIDEO)
                      INC A
              LD (VIDEO),A    ;VIDEO INDICATOR
              LD B(IY+2)
              LD E(IY+4)
              LD D,(IY+5)
              PUSH DE
              POP IX          ;RESET HISTOGRAM POINTER
              LD HL, SIGSTRT-1     ;RESET WAVEFORM POINTER
```

Introduction

```
            DEC C                  ;CHECK IF REQUIRED NUMBER OF CYCLES DONE
                                   ;IF SO, GOTO CONT3 AND STORE HISTOGRAM
            LD A,0
            CP C
            JR Z, CONT3
            LD DE, CONTINUE
            PUSH DE
            RET

CONT3       LD H,0                 ;STORE HISTOGRAM
            LD L,B
            ADD HL,HL
            EX DE,HL
            ADD IX,DE
            LD A,(IY+0)            ;STORE PARAMETERS
            LD (IX+0),A            ;NO OF CYCLES, 1 BYTE
            LD A,(IY+2)
            LD (IX+1),A            ;NO OF BINS, 1 BYTE
            LD A,(IY+8)
            LD (IX+2),A            ;TC GIVES PERIOD, 1 BYTE
            LD A,(IY+6)
            LD (IX+3),A
            LD A,(IY+7)
            LD (IX+4),A            ;GAIN, 2 BYTES
            LD A,0FFH
            LD (IX+5),A
            LD (IX+6),A
            LD (IX+7),A
            LD (IX+8),A            ;4 FFs TO MARK BLOCK END
            LD DE,9
            ADD IX,DE
            LD (TABLE+4),IX
            LD (ARG4),IX           ;ARG 4 STORE NEW BUFFER
                                   ;START
            LD A, 0E7H
            CP (IX+0)              ;TEST FOR END OF BUFFER
            JR Z, FINALL
            POP IY
            POP IX
            POP BC
            POP DE
            POP HL
            POP AF
            LD A,03                ;DISABLE AND STOP CTC
            OUT(CTC),A
            OUT(CTC+1),A
            OUT(CTC+2),A
            OUT(CTC+3),A
            RET
FINALL      POP IY                 ;IF END OF BUFFER GO HERE
            POP IX
            POP BC
            POP DE
            POP HL
            POP AF
            LD A,03
            OUT(CTC),A
            OUT(CTC+1),A
            OUT(CTC+2),A
            OUT(CTC+3),A
```

```
              POP HL
              LD HL,END
              PUSH HL
              RET
      END     DB 0DFH, 5BH      ;NASCOM SPECIFIC RESET INSTRUCTION
;END OF ROUTINE ACCESSED AT END OF EACH CYCLE AND SET OF CYCLES

;ROUTINE CALLED NEAR START TO INITIALIZE VARIABLES AND BUFFERS.

      SETUP   LD IY, TABLE      ;TRANSFERS ARGUMENTS TO TABLE OF VALUES
              LD HL, ARG2       ;ARG2
              LD A,(HL)
              LD (IY+0),A
              INC HL
              INC HL
              LD A,(HL)
              LD (IY+2),A
              INC HL
              INC HL
              LD A,(HL)
              LD (IY+4),A
              INC HL
              LD A,(HL)
              LD (IY+5),A
              INC HL
              LD A,(HL)
              LD (IY+6),A
              INC HL
              LD A,(HL)
              LD (IY+7),A
              LD DE,(ARG6)      ;ARG 6 IS THE TIME CONSTANT (TC) WHICH
                                ;DETERMINES BIN PERIOD
              LD A,E
              LD (IY+8),A
              LD A, 0FH         ;CONTROL BYTE WHICH SETS PIO PORTS TO OUTPUT
              OUT(GAIN+2),A
              OUT(GAIN+3),A
              OUT(DTOA+2),A
              OUT(DTOA+3),A
              LD C,(IY+0)
              LD B,(IY+2)
              LD A,(IY+6)       ;LIFTS LOW BYTE GAIN FROM TABLE
              OUT(GAIN),A       ;OUTPUTS TO D/A VIA PORT 18H
              LD A,(IY+7)
              OUT(GAIN+1),A     ;COMPLETES SETTING GAIN WITH 2ND BYTE
                                ;VIA 14 BIT D/A
              LD H,(IY+5)       ;PICK UP DATA START
              LD L,(IY+4)
              PUSH HL
              POP IX            ;GETS DATA START TO IX
              LD HL,SIGSTRT-1   ;HL REGISTER POINTS TO START OF SIGNAL
      GETTC   LD A,(IY+8)
              LD (TC+1),A       ;PUT TIME CONSTANT BYTE INTO PROGRAM

              PUSH HL
              PUSH DE
              PUSH BC
              PUSH IX
              POP HL            ;SAVE REGISTERS
```

Introduction

```
            ;NEXT PART INITIALIZES BUFFER TO ZEROS, AND LAST 230H BYTES TO 0E7H TO
            ;MARK THE END OF THE BUFFER.

                    LD   E,L
                    LD   D,H
                    LD   HL,DATFIN     ;END OF DATA AREA
                    OR   A
                    SBC  HL,DE
                    LD   C,L
                    LD   B,H
                    LD   H,D
                    LD   L,E
                    INC  DE
                    LD   (HL),0
                    LDIR
                    LD   HL, BUFFIN
                    LD   DE, BUFFIN+1
                    LD   BC, 230H
                    LD   (HL),0E7H
                    LDIR
                    POP  BC
                    POP  DE
                    POP  HL            ;RESTORE SOME REGISTERS
                    LD   (IY+10),0     ;DO FINAL SETTING UP
                    LD   A,57H
                    OUT  (CTC+1),A     ;COUNTS SPIKES
                    LD   A,0           ;TC OF 256
                    OUT  (CTC+1),A
                    RET

;END OF INITIALIZING ROUTINE


;START OF SPIKE COUNT ROUTINE
            SPIKCNT IN  A,(CTC+1)      ;SPIKE COUNT
                    NEG                ;CONVERTS FROM 256-COUNT TO ACTUAL COUNT
                    LD   E,A
                    SUB  (IY+10)       ;PREVIOUS COUNT SUBTRACTED
                    LD   (IY+10),E     ;NEW STORED COUNT
                    ADD  A,(IX+0)      ;SUM COUNTS FOR THAT BIN
                    LD   (IX+0),A
                    LD   A,0
                    ADC  A,(IX+1)      ;AND STORE SUM IN TWO BYTES
                    LD   (IX+1),A      ;IE HISTOGRAM BIN
                    INC  IX
                    INC  IX            ;IX POINTS TO NEXT BIN
                    RET

;FINISH OF SPIKE COUNT ROUTINE

;RESERVE SPACE AND SET ADDRESS OF TABLE WITH ASSEMBLER DIRECTIVE DS
            TABLE   DS  30
;TABLE OF INTERRUPT ADDRESSES FOR EACH CTC CHANNEL. NB ONLY CHANNEL 1 IS USED
            CALC1   ORG ((CALC1/8)*8)+8          ;VECTORS MUST START ON 5 BIT BOUNDARY
            VECAD   DW  PERBIN,PERBIN,PERBIN,PERBIN

;END OF ASSEMBLER PROGRAM.
```

Listing 2. Program to integrate EMGs and record start and finish times for bursts in an oscillated crab.

MUSCLE ANALYSIS PROGRAM ADAV.

```
0000                    ;ADAV  TO ANALYSE 2 CHANNELS MUSCLE
0000                    ;POTENTIALS AND NERVE SPIKES
0000                    ;3 BYTES PER TIME VERSION
0000                    ;SET UP CTC TO INTERRUPT
0000                    ; SPIKES GO TO 9
0000                    ;CH0 COUNTS DOWN EVERY 1 MSECS
0000                    ;CH2 (PORT A) IS USED TO CONTROL PRINT
0000                    ;OF MEMORY SEGMENT
0000                    ;CH 3  EVENTS
0000                    ;STORING INTERBURST TIMES, INTENSITIES
0000                    ; AND SPIKES
0000                    ;3 BYTES AND 2 BYTES FROM BUFF2ST
0000                    ;USES PORTS 4 AND 5 FOR 16 BIT D/A
                        ;SIGNAL FOR DC SERVO MOTOR. PORTS 18H AND 19H
                        ;CONTROL GAIN VIA 14 BIT D/A. PORT 14H
                        ;CONTROLS A/D MULTIPLEXOR. A/D START - PORT
                        ;14H BIT 4 . PORT 14H  BIT 5 IS SPARE BIT
                        ;;;;;;   FULLY INTERRUPTABLE     ;;;;;;

2B00            ORGIN   EQU 2B00H
2B00                    ORG ORGIN
2B00                    IDNT ORGIN, ORGIN
2B00
2B00 F3                 DI
2B01 F5                 PUSH AF
2B02 E5                 PUSH HL
2B03 C5                 PUSH BC
2B04 D5                 PUSH DE
2B05 DDE5               PUSH IX
2B07 FDE5               PUSH IY
2B09 D9                 EXX
2B0A E5                 PUSH HL
2B0B D9                 EXX
2B0C CD6C2E             CALL SETUP
2B0F 3E01       SETCTC  LD A, 01
2B11 D308               OUT (08),A
2B13 D309               OUT (09),A
2B15 D30A               OUT (10),A
2B17 D30B               OUT (11),A
2B19 D9                 EXX
2B1A 21212B             LD HL,CONT
2B1D E5                 PUSH HL
2B1E D9                 EXX
2B1F ED4D               RETI
2B21 3E2B       CONT    LD A, HIGH VECAD
2B23 ED47               LD I,A
2B25 ED5E               IM 2
2B27 3E68               LD A, LOW VECAD
2B29 D308               OUT (8),A
2B2B 3E95               LD A, 095H         ; INT./16,TIME
2B2D D308               OUT (8),A
2B2F 3ED5               LD A,0D5H
```

Introduction

```
2B31 D309              OUT (9),A        ;SET 9 TO INTERRUPT
2B33                                    ;EVERY SPIKE
2B33 3EA5              LD A, 0A5H       ;INT,TIME /256
2B35 D30A              OUT (10),A
2B37
2B37 3ED5              LD A, 0D5H       ;INT, COUNT
2B39 D30B              OUT (11),A
2B3B 3E01              LD A, 1          ;TC OF 1
2B3D D30B              OUT (11),A
2B3F FB                EI
2B40 00                NOP              ;HALT WAITS FOR EVENT PULSE
2B41 F3                DI
2B42 3EFA       TIMCON1 LD A,250        ;1MSEC
2B44 D308              OUT (8),A
2B46
2B46 3E01              LD A, 01
2B48 D309              OUT (9),A
2B4A
2B4A 3EFA       TIMCON2 LD A, 250       ;16 MSEC PRINT TIME
2B4C D30A              OUT (10),A
2B4E FB                EI
2B4F 3A4E2F     LOOP   LD A,(EXIT)
2B52 B7                OR A
2B53 28FA              JR Z,LOOP

2B55 F3                DI
2B56 D9                EXX
2B57 E1                POP HL
2B58 D9                EXX
2B59 FDE1              POP IY
2B5B DDE1              POP IX
2B5D D1                POP DE
2B5E C1                POP BC
2B5F E1                POP HL
2B60 F1                POP AF
2B61 C9                RET              ;TO PASCAL
2B68            CALC   ORG ((CALC/8)*8)+8
2B68 702B       VECAD  DEFW TIMES
2B6A EB2D              DEFW SPIKE
2B6C 442E              DEFW PRINT
2B6E F82D              DEFW PULSE

2B70 F3         TIMES  DI
2B71 F5                PUSH AF
2B72 E5                PUSH HL
2B73 D5                PUSH DE
2B74 C5                PUSH BC
2B75 CDF52E            CALL STAD0       ;SWITCH TO CH 0 AND
2B78                                    ;START CONVERSION
2B78 DB08              IN A, (08H)
2B7A D632              SUB 50
2B7C 6F                LD L,A
2B7D DB08       LOOP2  IN A, (08H)
2B7F BD                CP L
2B80 30FB              JR NC,LOOP2      ;WAITING FOR .1MSEC
2B82
2B82 214F2F            LD HL,TIME       ;INCREMENT TIME REG
2B85 34                INC (HL)
2B86 2006              JR NZ, FINTIM
2B88 23                INC HL
2B89 34                INC (HL)
2B8A 2002              JR NZ, FINTIM
```

```
2B8C 23                    INC HL
2B8D 34                    INC (HL)
2B8E DB15      FINTIM      IN A,(15H)        ;READ A/D
2B90 2A2F2F                LD HL,(BUFFCUR)
2B93 77                    LD (HL),A         ;STORES A/D
2B94 23                    INC HL
2B95 222F2F                LD (BUFFCUR),HL
2B98 CD0E2F                CALL STAD1        ;SWITCH TO CH 1
2B9B                                         ;AND START CONVERSION
2B9B 21522F                LD HL,T1          ;TEST FOR SPECIAL
2B9E                                         ;TIMES T1 TO T5
2B9E 3A502F                LD A,(TIME+1)
2BA1 BE                    CP (HL)
2BA2 200D                  JR NZ, CONT1
2BA4 3E30                  LD A,30H
2BA6 32D50B                LD (0BD5H),A
2BA9 3EFF                  LD A,255
2BAB D304                  OUT (4),A
2BAD 3EFF                  LD A,255
2BAF D305                  OUT (5),A
2BB1 21532F    CONT1       LD HL,T2
2BB4 3A502F                LD A,(TIME+1)
2BB7 BE                    CP (HL)
2BB8 200D                  JR NZ,CONT2
2BBA 3E31                  LD A,31H
2BBC 32D50B                LD (0BD5H),A
2BBF 3E00                  LD A,0
2BC1 D304                  OUT (4),A
2BC3 3E80                  LD A,128
2BC5 D305                  OUT (5),A
2BC7 21542F    CONT2       LD HL,T3
2BCA 3A502F                LD A,(TIME+1)
2BCD BE                    CP (HL)
2BCE 200D                  JR NZ, CONT3
2BD0 3E32                  LD A,32H
2BD2 32D50B                LD (0BD5H),A
2BD5 3E00                  LD A,0
2BD7 D304                  OUT (4),A
2BD9 3E00                  LD A,0
2BDB D305                  OUT (5),A
2BDD 21552F    CONT3       LD HL,T4
2BE0 3A502F                LD A,(TIME+1)
2BE3 BE                    CP (HL)
2BE4 200D                  JR NZ, CONT4
2BE6 3E33                  LD A,33H
2BE8 32D50B                LD (0BD5H),A
2BEB 3E00                  LD A,0
2BED D304                  OUT (4),A
2BEF 3E80                  LD A,128
2BF1 D305                  OUT (5),A
2BF3 21562F    CONT4       LD HL,T5
2BF6 3A502F                LD A,(TIME+1)
2BF9 BE                    CP (HL)
2BFA 206B                  JR NZ, CONT5
2BFC 3E34                  LD A,34H
2BFE 32D50B                LD (0BD5H),A
2C01 214F2F                LD HL, TIME
2C04 3E00                  LD A,0
2C06 77                    LD (HL),A
2C07 23                    INC HL
2C08 77                    LD (HL),A
2C09 23                    INC HL
```

Introduction

```
2C0A  77              LD (HL),A         ;ZERO TIME
2C0B  ED5B412F        LD DE,(OSCNOCUR)
2C0F  13              INC DE
2C10  ED53412F        LD (OSCNOCUR),DE
2C14  3E30            LD A,30H
2C16  83              ADD A,E
2C17  32E80B          LD (0BE8H),A
2C1A  7B              LD A,E
2C1B
2C1B  2A372F          LD HL,(DATCUR)
2C1E  54              LD D,H
2C1F  5D              LD E,L
2C20  13              INC DE
2C21  77              LD (HL),A
2C22  010700          LD BC,7
2C25  EDB0            LDIR
2C27  22372F          LD (DATCUR),HL
2C2A  2A392F          LD HL,(DATCUR1)
2C2D  54              LD D,H
2C2E  5D              LD E,L
2C2F  13              INC DE
2C30  77              LD (HL),A
2C31  010700          LD BC,7
2C34  EDB0            LDIR
2C36  22392F          LD (DATCUR1),HL
2C39  5F              LD E,A
2C3A  1600            LD D,0
2C3C  2A3D2F          LD HL,(SPKCUR)   ;MARK END OF CYCLE
2C3F  36FF            LD (HL),0FFH
2C41  23              INC HL
2C42  77              LD (HL),A
2C43  23              INC HL
2C44  223D2F          LD (SPKCUR),HL
2C47  2A3F2F          LD HL,(OSCNOST)
2C4A  AF              XOR A
2C4B  ED52            SBC HL,DE
2C4D  7D              LD A,L
2C4E  B4              OR H
2C4F  2016            JR NZ,CONT5
2C51  EF       FIN0   DB 0EFH
2C52  434F554E        DEFM "COUNT UP"
2C56  54205550
2C5A  00              DB 0
2C5B  C1              POP BC            ;ALL FINISHED
2C5C  D1              POP DE
2C5D  E1              POP HL
2C5E  F1              POP AF
2C5F  3E01            LD A,1
2C61  324E2F          LD (EXIT),A
2C64  F3              DI
2C65  ED4D            RETI

2C67  DB08     CONT5  IN A,(08H)
2C69  D632            SUB 50
2C6B  6F              LD L,A
2C6C  DB08     LOOP1  IN A,(08H)
2C6E  BD              CP L
2C6F  30FB            JR NC,LOOP1       ;WAITING FOR .1MSEC

2C71  DB15            IN A,(15H)        ;READ A/D CH 1
2C73  2A2F2F          LD HL,(BUFFCUR)   ;AND STORE
2C76  77              LD (HL),A
```

```
2C77 23                    INC HL
2C78 222F2F                LD (BUFFCUR),HL
2C7B 23                    INC HL
2C7C 3E50                  LD A,50H         ;TEST FOR END OF BUFFER
2C7E BC                    CP H
2C7F 2006                  JR NZ,CONT6
2C81 2A2B2F                LD HL,(BUFFST)
2C84 222F2F                LD (BUFFCUR),HL  ;RESETS A/D BUFFER
2C87 3A312F     CONT6      LD A,(BURST0)
2C8A FE00                  CP 0
2C8C 2877                  JR Z,CONT7       ;CHECK IF IN BURST

2C8E                       ;IN BURST - MEASURE INTENSITY
2C8E 2A2F2F                LD HL,(BUFFCUR)
2C91 2B                    DEC HL
2C92 2B                    DEC HL
2C93 3A4C2F                LD A,(THOLD)     ;GETS CH 0 A/D VALUE
2C96 BE                    CP (HL)
2C97 300C                  JR NC, CONT9
2C99 3A502F                LD A,(TIME+1)    ;ABOVE THRESHOLD
2C9C 32482F                LD (TEMPTM+1),A
2C9F 3A4F2F                LD A,(TIME)
2CA2 32472F                LD (TEMPTM),A    ;STORE TIME, LAST ABOVE
2CA5                                        ;THRESHOLD EVENT IN TEMPTM
2CA5 7E         CONT9      LD A,(HL)
2CA6 2A432F                LD HL,(INTENS)
2CA9 5F                    LD E,A
2CAA 1600                  LD D,0
2CAC 19                    ADD HL,DE
2CAD 22432F                LD (INTENS),HL   ;INTEGRATE ACTIVITY
2CB0                       ;CHECK FOR END OF BURST
2CB0 ED5B472F              LD DE,(TEMPTM)
2CB4 2A4F2F                LD HL,(TIME)
2CB7 AF                    XOR A
2CB8 ED52                  SBC HL,DE
2CBA 3A4B2F                LD A,(DELAY)
2CBD BD                    CP L
2CBE 3076                  JR NC, FIN1
2CC0 2A372F                LD HL,(DATCUR)   ;DELAY ABOVE SET LEVEL
2CC3 3A472F                LD A,(TEMPTM)    ;STORE IN DATCUR WITH
2CC6 77                    LD (HL),A        ;LSB SET TO 0
2CC7 23                    INC HL
2CC8 3A482F                LD A,(TEMPTM+1)
2CCB 77                    LD (HL),A
2CCC 23                    INC HL
2CCD 3A432F                LD A,(INTENS)
2CD0 77                    LD (HL),A
2CD1 23                    INC HL
2CD2 3A442F                LD A,(INTENS+1)
2CD5 77                    LD (HL),A
2CD6 23                    INC HL
2CD7 22372F                LD (DATCUR),HL   ;STORE INTEGRAL AND
2CDA                                        ;UPDATE POINTER
2CDA 210000                LD HL,0
2CDD 22432F                LD (INTENS),HL
2CE0 2A372F                LD HL,(DATCUR)
2CE3 23                    INC HL
2CE4 3E60                  LD A,60H
2CE6 BC                    CP H
2CE7 2010                  JR NZ,CONT14
2CE9 EF                    DB 0EFH
2CEA 44415443              DEFM "DATCUR FULL"
```

Introduction

```
2CEE  55522046
2CF2  554C4C
2CF5  00                      DB 0
2CF6  C3512C                  JP FIN0
2CF9  3E00         CONT14     LD A,0
2CFB  32312F                  LD (BURST0),A    ;SIGNIFY OUT OF BURST
2CFE  3E20                    LD A,20H
2D00  32E00B                  LD (0BE0H),A
2D03  1831                    JR FIN1
                   ;;;;;;;;NOT IN BURST;;;;;;;;;

2D05  2A2F2F       CONT7      LD HL,(BUFFCUR)
2D08  2B                      DEC HL
2D09  2B                      DEC HL
2D0A  3A4C2F                  LD A,(THOLD)            ;THRESHOLD VALUE
2D0D  BE                      CP (HL)
2D0E  3026                    JR NC,FIN1
2D10  3E01                    LD A,1
2D12  32312F                  LD (BURST0),A    ;SIGNIFIES BURST
2D15  3EFF                    LD A,0FFH
2D17  32E00B                  LD (0BE0H),A
2D1A  214F2F                  LD HL, TIME
2D1D  010300                  LD BC,3
2D20  ED5B372F                LD DE,(DATCUR)
2D24  EDB0                    LDIR
2D26  ED53372F                LD (DATCUR),DE
2D2A  1B                      DEC DE
2D2B  1B                      DEC DE
2D2C  1A                      LD A,(DE)
2D2D  32482F                  LD (TEMPTM+1),A  ;STORE TIME
2D30  1B                      DEC DE           ;POINT TO LSB TIME
2D31  1A                      LD A,(DE)
2D32  32472F                  LD (TEMPTM),A
2D35  12                      LD (DE),A

                   ;;;;;;;;;;;;;SECOND CHANNEL;;;;;;;;
2D36  3A322F       FIN1       LD A,(BURST1)
2D39  FE00                    CP 0
2D3B  2877                    JR Z,CONT10      ;CHECK IF IN BURST

2D3D                          ;IN BURST - MEASURE INTENSITY
2D3D  2A2F2F                  LD HL,(BUFFCUR)
2D40  2B                      DEC HL
2D41  3A4D2F                  LD A,(THOLD1)    ;GETS CH 1 THRESHOLD
2D44                                           ;AND COMPARES A/D VALUE
2D44  BE                      CP (HL)
2D45  300C                    JR NC, CONT11
2D47  3A502F                  LD A,(TIME+1)
2D4A  324A2F                  LD (TEMPTM1+1),A
2D4D  3A4F2F                  LD A,(TIME)
2D50  32492F                  LD (TEMPTM1),A   ;STORE TIME, LAST ABOVE
2D53                                           ;THRESHOLD EVENT IN TEMPTM1
2D53  7E           CONT11     LD A,(HL)
2D54  2A452F                  LD HL,(INTENS1)
2D57  5F                      LD E,A
2D58  1600                    LD D,0
2D5A  19                      ADD HL,DE
2D5B  22452F                  LD (INTENS1),HL  ;INTEGRATE ACTIVITY

2D5E                          ;CHECK FOR END OF BURST
2D5E  ED5B492F                LD DE,(TEMPTM1)
2D62  2A4F2F                  LD HL,(TIME)
```

```
2D65  AF                   XOR A
2D66  ED52                 SBC HL,DE
2D68  3A4B2F               LD A, (DELAY)
2D6B  BD                   CP L
2D6C  3076                 JR NC, FIN2
2D6E  2A392F               LD HL, (DATCUR1)
2D71  3A492F               LD A, (TEMPTM1)
2D74  77                   LD (HL),A
2D75  23                   INC HL
2D76  3A4A2F               LD A, (TEMPTM1+1)
2D79  77                   LD (HL),A
2D7A  23                   INC HL
2D7B  3A452F               LD A,(INTENS1)
2D7E  77                   LD (HL),A
2D7F  23                   INC HL
2D80  3A462F               LD A,(INTENS1+1)
2D83  77                   LD (HL),A
2D84  23                   INC HL
2D85  22392F               LD (DATCUR1),HL
2D88  210000               LD HL,0
2D8B  22452F               LD (INTENS1),HL
2D8E  2A392F               LD HL, (DATCUR1)
2D91  23                   INC HL
2D92  3E70                 LD A,70H
2D94  BC                   CP H
2D95  2011                 JR NZ,CONT15
2D97  EF                   DB 0EFH
2D98  44415443             DEFM "DATCUR1 FULL"
2D9C  55523120
2DA0  46554C4C
2DA4  00                   DB 0
2DA5  C3512C               JP FIN0
2DA8  3E00        CONT15   LD A,0
2DAA  32322F               LD (BURST1),A
2DAD  3E20                 LD A,20H
2DAF  32E20B               LD (0BE2H),A
2DB2  1830                 JR FIN2

            ;;;;;;;;NOT IN BURST;;;;;;;;;

2DB4  2A2F2F      CONT10   LD HL,(BUFFCUR)
2DB7  2B                   DEC HL
2DB8  3A4D2F               LD A,(THOLD1)          ;THRESHOLD VALUE
2DBB  BE                   CP (HL)
2DBC  3026                 JR NC,FIN2
2DBE  3E01                 LD A,1
2DC0  32322F               LD (BURST1),A    ;SIGNIFIES BURST
2DC3  3EFF                 LD A,0FFH
2DC5  32E20B               LD (0BE2H),A
2DC8  214F2F               LD HL, TIME
2DCB  010300               LD BC,3
2DCE  ED5B392F             LD DE, (DATCUR1)
2DD2  EDB0                 LDIR
2DD4  ED53392F             LD (DATCUR1),DE
2DD8  1B                   DEC DE
2DD9  1B                   DEC DE
2DDA  1A                   LD A,(DE)
2DDB  324A2F               LD (TEMPTM1+1),A       ;STORE TIME
2DDE  1B                   DEC DE           ;POINT TO LSB TIME
2DDF  1A                   LD A,(DE)
2DE0  32492F               LD (TEMPTM1),A
2DE3  12                   LD (DE),A
```

Introduction

```
                        ;;;;;;;;;;;;;;;;;
2DE4  C1        FIN2    POP BC
2DE5  D1                POP DE
2DE6  E1                POP HL
2DE7  F1                POP AF
2DE8  FB                EI
2DE9  ED4D              RETI
2DEB  E5        SPIKE   PUSH HL
2DEC  21EC0B            LD HL,0BECH
2DEF  34                INC (HL)
2DF0  2A3D2F            LD HL,(SPKCUR)
2DF3  34                INC (HL)
2DF4  E1                POP HL
2DF5  FB                EI
2DF6  ED4D              RETI
2DF8  FB        PULSE   EI
2DF9  ED4D              RETI            ;DISCONNECTS PULSE
2DFB  F5                PUSH AF
2DFC  E5                PUSH HL
2DFD  D5                PUSH DE
2DFE  3ADA0B            LD A,(0BDAH)
2E01  3C                INC A
2E02  32DA0B            LD (0BDAH),A
2E05  3E00              LD A,0
2E07  214F2F            LD HL, TIME
2E0A  77                LD (HL),A
2E0B  23                INC HL
2E0C  77                LD (HL),A
2E0D  23                INC HL
2E0E  77                LD (HL),A
2E0F  ED5B412F          LD DE,(OSCNOCUR)
2E13  13                INC DE
2E14  ED53412F          LD (OSCNOCUR),DE
2E18  2A3F2F            LD HL,(OSCNOST)
2E1B  AF                XOR A
2E1C  ED52              SBC HL,DE
2E1E  7D                LD A,L
2E1F  B4                OR H
2E20  201C              JR NZ, CONT12
2E22  3E01              LD A,1
2E24  324E2F            LD (EXIT),A
2E27  EF                DB 0EFH
2E28  50554C53          DEFM "PULSES FINISHED"
2E2C  45532046
2E30  494E4953
2E34  484544
2E37  00                DB 00
2E38  D1                POP DE
2E39  E1                POP HL
2E3A  F1                POP AF
2E3B  F3                DI
2E3C  ED4D              RETI
2E3E  D1        CONT12  POP DE
2E3F  E1                POP HL
2E40  F1                POP AF
2E41  FB                EI
2E42  ED4D              RETI

2E44            PRINT   ;;; INCREMENTS SPIKE POINTER ;;;
2E44  F5                PUSH AF
2E45  E5                PUSH HL
2E46  2A3D2F            LD HL,(SPKCUR)
```

```
2E49 23                    INC HL
2E4A 223D2F                LD (SPKCUR),HL
2E4D 3E7F                  LD A,7FH
2E4F BC                    CP H
2E50 200B                  JR NZ,CONT16
2E52 2A3B2F                LD HL,(SPKST)
2E55 223D2F                LD (SPKCUR),HL
2E58 E1                    POP HL
2E59 F1                    POP AF
2E5A FB                    EI
2E5B ED4D                  RETI
2E5D 3AEA0B        CONT16  LD A,(0BEAH)
2E60 3C                    INC A
2E61 32EA0B                LD (0BEAH),A
2E64 E1                    POP HL
2E65 F1                    POP AF
2E66 FB                    EI
2E67 ED4D                  RETI
2E69 FB                    EI
2E6A ED4D                  RETI
2E6C 3E0F          SETUP   LD A,0FH
2E6E D316                  OUT(16H),A
2E70 D306                  OUT (6),A
2E72 D307                  OUT (7),A
2E74 D31A                  OUT (1AH),A
2E76 D31B                  OUT (1BH),A    ;D/A PORTS TO OUTPUT
2E78 3E4F                  LD A,4FH
2E7A D317                  OUT(17H),A     ;SETS UP MULTIPLEX /A/D
2E7C 3E00                  LD A,0
2E7E D318                  OUT (18H),A
2E80 3E20                  LD A,32
2E82 D319                  OUT (19H),A    ;GAIN
2E84 3E00                  LD A,0
2E86 D304                  OUT (4),A
2E88 3E80                  LD A,128
2E8A D305                  OUT (5),A
2E8C 2A2B2F                LD HL,(BUFFST)
2E8F 222F2F                LD (BUFFCUR),HL
2E92 2A3B2F                LD HL,(SPKST)
2E95 223D2F                LD (SPKCUR),HL
2E98 3E00                  LD A,0
2E9A 324E2F                LD (EXIT),A
2E9D 32312F                LD (BURST0),A
2EA0 32322F                LD (BURST1),A
2EA3 32512F                LD (TIME+2),A
2EA6 2A332F                LD HL,(DATST)
2EA9 22372F                LD (DATCUR),HL
2EAC 2A352F                LD HL,(DATST1)
2EAF 22392F                LD (DATCUR1),HL
2EB2 210000                LD HL,0
2EB5 224F2F                LD (TIME),HL
2EB8 22412F                LD (OSCNOCUR),HL
2EBB 22432F                LD (INTENS),HL
2EBE 22452F                LD (INTENS1),HL
2EC1 2A2B2F                LD HL,(BUFFST)
2EC4 54                    LD D,H
2EC5 5D                    LD E,L
2EC6 13                    INC DE
2EC7 010040                LD BC,4000H
2ECA 3600                  LD (HL),0
2ECC EDB0                  LDIR
2ECE 3A412F                LD A,(OSCNOCUR)
```

Introduction

```
2ED1  2A372F           LD HL,(DATCUR)
2ED4  54               LD D,H
2ED5  5D               LD E,L
2ED6  13               INC DE
2ED7  77               LD (HL),A
2ED8  010700           LD BC,7
2EDB  EDB0             LDIR
2EDD  22372F           LD (DATCUR),HL
2EE0  2A392F           LD HL,(DATCUR1)
2EE3  54               LD D,H
2EE4  5D               LD E,L
2EE5  13               INC DE
2EE6  77               LD (HL),A
2EE7  010700           LD BC,7
2EEA  EDB0             LDIR
2EEC  22392F           LD (DATCUR1),HL

2EEF  C630             ADD A,30H
2EF1  32E80B           LD (0BE8H),A
2EF4  C9               RET
2EF5  3E00      STAD0  LD A,0
2EF7  D314             OUT (14H),A
2EF9  3E30             LD A,30H
2EFB  32CE0B           LD (0BCEH),A
2EFE  3E10             LD A,10H         ;SWITCH TO 0 + START PULSE
2F00  D314             OUT (14H),A
2F02  3AD00B           LD A,(0BD0H)
2F05  3C               INC A
2F06  32D00B           LD (0BD0H),A
2F09  3E00             LD A,0
2F0B  D314             OUT (14H),A
2F0D  C9               RET
2F0E  3E01      STAD1  LD A,1
2F10  D314             OUT (14H),A
2F12  3E31             LD A,31H
2F14  32CF0B           LD (0BCFH),A
2F17  3E11             LD A,11H         ;SWITCH TO 1
2F19  D314             OUT (14H),A
2F1B  3AD20B           LD A,(0BD2H)
2F1E  3C               INC A
2F1F  32D20B           LD (0BD2H),A
2F22  3E01             LD A,1
2F24  D314             OUT (14H),A
2F26  C9               RET
2F27  00        TABLE  NOP
2F28  00               NOP
2F29  00               NOP
2F2A  00               NOP
2F2B  0031      BUFFST  DW 3100H
2F2D  001F      BUFFLEN DW 1F00H
2F2F  0031      BUFFCUR DW 3100H
2F31  00        BURST0  DB 0
2F32  00        BURST1  DB 0
2F33  0050      DATST   DW 5000H
2F35  0060      DATST1  DW 6000H
2F37  0050      DATCUR  DW 5000H
2F39  0060      DATCUR1 DW 6000H
2F3B  0070      SPKST   DW 7000H
2F3D  0070      SPKCUR  DW 7000H
2F3F  0500      OSCNOST DW 5
2F41  0000      OSCNOCUR DW 0
```

P.J.Fraser

```
2F43 0000              INTENS   DW 0
2F45 0000              INTENS1  DW 0
2F47 0000              TEMPTM   DW 0
2F49 0000              TEMPTM1  DW 0
2F4B 64                DELAY    DB 100
2F4C 28                THOLD    DB 40
2F4D 28                THOLD1   DB 40
2F4E 00                EXIT     DB 0
2F4F 000000            TIME     DB 0,0,0      ;3 BYTES LOW FIRST
2F52 01                T1       DB 1
2F53 04                T2       DB 4
2F54 29                T3       DB 41
2F55 2C                T4       DB 44
2F56 51                T5       DB 81
                       ;; ROUTINE TO CHECK TRIGGER LEVELS ;;;
                       ;;    CHANNEL 0    ;;
2F57 F3                ADTRIG0  DI
2F58 F5                         PUSH AF
2F59 E5                         PUSH HL
2F5A C5                         PUSH BC
2F5B D5                         PUSH DE
2F5C DDE5                       PUSH IX
2F5E FDE5                       PUSH IY
2F60 D9                         EXX
2F61 E5                         PUSH HL
2F62 D9                         EXX
2F63
2F63 3A4C2F                     LD A,(THOLD)
2F66 DF68                        DB 0DFH,68H
2F68 214C2F           LOOP10    LD HL,THOLD
2F6B CDF52E                     CALL STAD0
2F6E 3E10                       LD A,10H
2F70 FF                         DB 0FFH           ;DELAY
2F71 DB15                       IN A,(15H)
2F73 BE                         CP (HL)
2F74 3807                       JR C,CONT20
2F76 3EFF                       LD A,0FFH
2F78 32CC0B                     LD (0BCCH),A
2F7B 1805                       JR CONT23
2F7D 3E20             CONT20    LD A,20H
2F7F 32CC0B                     LD (0BCCH),A
2F82 DF62             CONT23    DB 0DFH,62H       ;IN SCANS FOR INPUT
2F84 30E2                       JR NC,LOOP10      ;NO INPUT
2F86                            ;;  INPUT;;
2F86 FE13                       CP 13H            ;UP ARROW
2F88 2010                       JR NZ,CONT21
2F8A 3A4C2F                     LD A,(THOLD)
2F8D 3C                         INC A
2F8E 324C2F                     LD (THOLD),A
2F91 DF68DF6A                   DB 0DFH,68H,0DFH,6AH
2F95 3E13                       LD A,13H
2F97 F7                         DB 0F7H
2F98 18CE                       JR LOOP10
2F9A FE14             CONT21    CP 14H
2F9C 2010                       JR NZ,CONT22
2F9E 3A4C2F                     LD A,(THOLD)
2FA1 3D                         DEC A
2FA2 324C2F                     LD (THOLD),A
2FA5 DF68DF6A                   DB 0DFH,68H,0DFH,6AH
2FA9 3E13                       LD A,13H
2FAB F7                         DB 0F7H
```

Introduction

```
2FAC 18BA              JR LOOP10
2FAE FE46      CONT22  CP 46H           ;ASCII F
2FB0 20B6              JR NZ,LOOP10
2FB2 F3                DI
2FB3 D9                EXX
2FB4 E1                POP HL
2FB5 D9                EXX
2FB6 FDE1              POP IY
2FB8 DDE1              POP IX
2FBA D1                POP DE
2FBB C1                POP BC
2FBC E1                POP HL
2FBD F1                POP AF

2FBE C9                RET
                ;; CHANNEL 1   ;;
2FBF F3        ADTRIG1 DI
2FC0 F5                PUSH AF
2FC1 E5                PUSH HL
2FC2 C5                PUSH BC
2FC3 D5                PUSH DE
2FC4 DDE5              PUSH IX
2FC6 FDE5              PUSH IY
2FC8 D9                EXX
2FC9 E5                PUSH HL
2FCA D9                EXX

2FCB 3A4D2F            LD A,(THOLD1)
2FCE DF68              DB 0DFH,68H
2FD0 214D2F    LOOP11  LD HL,THOLD1
2FD3 CD0E2F            CALL STAD1
2FD6 3E10              LD A,10H
2FD8 FF                DB 0FFH          ;DELAY
2FD9 DB15              IN A,(15H)
2FDB BE                CP (HL)
2FDC 3807              JR C,CONT25
2FDE 3EFF              LD A,0FFH
2FE0 32CC0B            LD (0BCCH),A
2FE3 1805              JR CONT28
2FE5 3E20      CONT25  LD A,20H
2FE7 32CC0B            LD (0BCCH),A
2FEA DF62      CONT28  DB 0DFH,62H     ;IN SCANS FOR INPUT
2FEC 30E2              JR NC,LOOP11    ;NO INPUT
2FEE                   ;; INPUT;;
2FEE FE13              CP 13H           ;UP ARROW
2FF0 2010              JR NZ,CONT26
2FF2 3A4D2F            LD A,(THOLD1)
2FF5 3C                INC A
2FF6 324D2F            LD (THOLD1),A
2FF9 DF68DF6A          DB 0DFH,68H,0DFH,6AH
2FFD 3E13              LD A,13H
2FFF F7                DB 0F7H
3000 18CE              JR LOOP11
3002 FE14      CONT26  CP 14H
3004 2010              JR NZ,CONT27
3006 3A4D2F            LD A,(THOLD1)
3009 3D                DEC A
300A 324D2F            LD (THOLD1),A
300D DF68DF6A          DB 0DFH,68H,0DFH,6AH
3011 3E13              LD A,13H
3013 F7                DB 0F7H
3014 18BA              JR LOOP11
```

```
3016 FE46      CONT27    CP  46H              ;ASCII F
3018 20B6                JR  NZ,LOOP11
301A F3                  DI
301B D9                  EXX
301C E1                  POP HL
301D D9                  EXX
301E FDE1                POP IY
3020 DDE1                POP IX
3022 D1                  POP DE
3023 C1                  POP BC
3024 E1                  POP HL
3025 F1                  POP AF

3026 C9                  RET


3027 F3        DELAY1    DI
3028 F5                  PUSH AF
3029 E5                  PUSH HL
302A D5                  PUSH DE
302B 3E01                LD  A,01
302D D308                OUT (8),A
302F 3E25                LD  A, 25H           ;/256 NO INTERRUPT
3031 D308                OUT (8),A
3033 3EFA                LD  A,250            ;8 MSEC
3035 D308                OUT (8),A
3037 210000              LD  HL,0
303A 227D30              LD  (SEC),HL
303D 227B30              LD  (MSEC),HL
3040 DB08     LOOPD      IN  A, (08)
3042 FEFA                CP  250
3044 20FA                JR  NZ,LOOPD         ;EVERY 1MSEC
3046 2A7B30              LD  HL,(MSEC)
3049 23                  INC HL
304A 227B30              LD  (MSEC),HL
304D 117D00              LD  DE,125           ;1000 MSEC
3050 AF                  XOR A
3051 ED52                SBC HL,DE
3053 7D                  LD  A,L
3054 B4                  OR  H
3055 20E9                JR  NZ, LOOPD        ;EVERY 1 SECOND
3057 227B30              LD  (MSEC),HL        ;ZERO MSEC
305A 2A7D30              LD  HL,(SEC)
305D 23                  INC HL
305E 227D30              LD  (SEC),HL
3061 3AE40B              LD  A, (0BE4H)       ;VIDEO MARKER
3064 3C                  INC A
3065 32E40B              LD  (0BE4H),A
3068 ED5B7F30            LD  DE,(DELSEC)
306C AF                  XOR A
306D ED52                SBC HL,DE
306F 7D                  LD  A,L
3070 B4                  OR  H
3071 20CD                JR  NZ,LOOPD
3073 3E01                LD  A,01
3075 D308                OUT (8),A            ;DISABLE CTC
3077 D1                  POP DE
3078 E1                  POP HL
3079 F1                  POP AF
307A C9                  RET                  ;END OF DELAY
307B 0000     MSEC       DW  0
307D 0000     SEC        DW  0
```

Introduction

```
307F  8403              DELSEC  DW 900      ;15 MINUTES

3081  F5        STORE1  PUSH AF
3082  E5                PUSH HL
3083  D5                PUSH DE
3084  C5                PUSH BC
3085  2A372F            LD HL,(DATCUR)
3088  ED5B332F          LD DE, (DATST)
308C  AF                XOR A
308D  ED52              SBC HL,DE
308F  44                LD B,H
3090  4D                LD C,L
3091  EB                EX DE,HL
3092  ED5BF230          LD DE,(STORECUR)
3096  EDB0              LDIR
3098  ED53F230          LD (STORECUR),DE
309C  2A392F            LD HL,(DATCUR1)
309F  ED5B352F          LD DE, (DATST1)
30A3  AF                XOR A
30A4  ED52              SBC HL,DE
30A6  44                LD B,H
30A7  4D                LD C,L
30A8  EB                EX DE,HL
30A9  ED5BF430          LD DE,(STORECUR1)
30AD  EDB0              LDIR
30AF  ED53F430          LD (STORECUR1),DE
30B3  C1                POP BC
30B4  D1                POP DE
30B5  E1                POP HL
30B6  F1                POP AF
30B7  C9                RET

30B8  E5      INITSPACE1    PUSH HL
30B9  2AEE30            LD HL,(STOREST)
30BC  22F230            LD (STORECUR),HL
30BF  2AF030            LD HL,(STOREST1)
30C2  22F430            LD (STORECUR1),HL
30C5  E1                POP HL
30C6  C9                RET
30C7
30C7  F5       WRITE1   PUSH AF
30C8  E5                PUSH HL
30C9  D5                PUSH DE
30CA  C5                PUSH BC
30CB  DDE5              PUSH IX
30CD  FDE5              PUSH IY
30CF  2AEE30            LD HL, (STOREST)
30D2  111000            LD DE,10H        ;INCLUDE ID BLOCK
30D5  AF                XOR A
30D6  ED52              SBC HL,DE
30D8  220C0C            LD (0C0CH),HL
30DB  ED5BF430          LD DE, (STORECUR1)
30DF  ED530E0C          LD (0C0EH),DE
30E3  DF                RST 18H
30E4  57                DB "W"
30E5  FDE1              POP IY
30E7  DDE1              POP IX
30E9  C1                POP BC
30EA  D1                POP DE
30EB  E1                POP HL
30EC  F1                POP AF
```

```
30ED  C9                    RET
30EE

30EE  0080        STOREST   DW  8000H
30F0  00A0        STOREST1  DW  0A000H
30F2  0080        STORECUR  DW  8000H
30F4  00A0        STORECUR1 DW  0A000H
```

```
ADTRIG0    2F57    ADTRIG1    2FBF    BUFFCUR   2F2F
BUFFLEN    2F2D    BUFFST     2F2B    BURST0    2F31
BURST1     2F32    CALC       2B68    CONT      2B21
CONT1      2BB1    CONT10     2DB4    CONT11    2D53
CONT12     2E3E    CONT14     2CF9    CONT15    2DA8
CONT16     2E5D    CONT2      2BC7    CONT20    2F7D
CONT21     2F9A    CONT22     2FAE    CONT23    2F82
CONT25     2FE5    CONT26     3002    CONT27    3016
CONT28     2FEA    CONT3      2BDD    CONT4     2BF3
CONT5      2C67    CONT6      2C87    CONT7     2D05
CONT9      2CA5    DATCUR     2F37    DATCUR1   2F39
DATST      2F33    DATST1     2F35    DELAY     2F4B
DELAY1     3027    DELSEC     307F    EXIT      2F4E
FIN0       2C51    FIN1       2D36    FIN2      2DE4
FINTIM     2B8E    INITSPACE1 30B8    INTENS    2F43
INTENS1    2F45    LOOP       2B4F    LOOP1     2C6C
LOOP10     2F68    LOOP11     2FD0    LOOP2     2B7D
LOOPD      3040    MSEC       307B    ORGIN     2B00
OSCNOCUR   2F41    OSCNOST    2F3F    PRINT     2E44
PULSE      2DF8    SEC        307D    SETCTC    2B0F
SETUP      2E6C    SPIKE      2DEB    SPKCUR    2F3D
SPKST      2F3B    STAD0      2EF5    STAD1     2F0E
STORE1     3081    STORECUR   30F2    STORECUR1 30F4
STOREST    30EE    STOREST1   30F0    T1        2F52
T2         2F53    T3         2F54    T4        2F55
T5         2F56    TABLE      2F27    TEMPTM    2F47
TEMPTM1    2F49    THOLD      2F4C    THOLD1    2F4D
TIMCON1    2B42    TIMCON2    2B4A    TIME      2F4F
TIMES      2B70    VECAD      2B68    WRITE1    30C7
```

Listing 3. Pascal program used to manipulate the machine code routines used in *Listing 2.*

```
PROGRAM SWCBAN10;
LABEL MENU ;
CONST
    ORGIN=$2B00;
    ADTRIG0=$2F57;
    ADTRIG1=$2FBF;
    DELAY1=$3027;
    DELSEC=$307F;
    DATST=$2F33;
    DATST1=$2F35;
    DATCUR=$2F37;
    DATCUR1=$2F39;
    STOREST=$30EE;
    STOREST1=$30F0;
    STORECUR=$30F2;
    STORECUR1=$30F4;
    STORE1=$3081;
    WRITE1=$30C7;
    INITSPACE1=$30B8;
```

Introduction

```
VAR S: STRING[1];
    DATE: STRING[8];
    N,I,J,K: INTEGER;

PROCEDURE OSCILLATE;
  EXTERNAL ORGIN;
PROCEDURE TRIG0;
  EXTERNAL ADTRIG0;
PROCEDURE TRIG1;
  EXTERNAL ADTRIG1;
PROCEDURE DELAYCD;
  EXTERNAL DELAY1;

PROCEDURE DELAY(SEC: INTEGER);
  BEGIN
    MEM[DELSEC]:= SEC-(SEC DIV 256) ;
    MEM[DELSEC+1]:=SEC DIV 256 ;
    DELAYCD;
  END;


PROCEDURE SETTRIG0;
 BEGIN
  WRITELN('SET A/D CHANNEL 0 TRIGGER LEVEL');
  TRIG0;
 END;
PROCEDURE SETTRIG1;
 BEGIN
  WRITELN('SET A/D CHANNEL 1 TRIGGER LEVEL');
  TRIG1;
 END;
PROCEDURE SETANGLE(ANGLE: INTEGER);
  BEGIN
    OUT($1F,15);
    OUT($1D,127);
    WRITELN(ANGLE);
    IF ANGLE=-45 THEN OUT($1D,127);
    IF ANGLE=0 THEN OUT($1D,165);
    IF ANGLE=45 THEN OUT($1D,202);
    DELAY(30);
  END;

PROCEDURE STORE;
  EXTERNAL STORE1;
PROCEDURE WRIT;
  EXTERNAL WRITE1;
PROCEDURE INITSPACE;
  EXTERNAL INITSPACE1;

PROCEDURE ID;
      BEGIN
        FOR J:=0 TO 7 DO
            BEGIN
            MEM[K+J]:=ORD(MID(DATE,J+1,1));
            END;
        MEM[K+8]:=I;
        MEM[K+9]:=N;
      END;

PROCEDURE NOTROOM;
  BEGIN
```

```
            ID;
            WRIT;
            INITSPACE;
            STORE;
            N:=0;
        END;
   PROCEDURE CALCDATASPACE;
     VAR S1,S2,L1,L2: INTEGER;
     BEGIN

   S1:=MEM[DATCUR]-MEM[DATST]+(MEM[DATCUR+1]-MEM[DATST+1])*256;
   S2:=MEM[DATCUR1]-MEM[DATST1]+(MEM[DATCUR1+1]-MEM[DATST1+1])*256;
   L1:=MEM[STORECUR]-MEM[STOREST]+(MEM[STORECUR+1]-MEM[STOREST+1])*256;
   L2:=MEM[STORECUR1]-MEM[STOREST1]+(MEM[STORECUR1+1]-MEM[STOREST1+1])*256;
   IF S1+L1<$2000 THEN IF S2+L2<$2000 THEN STORE ELSE NOTROOM;

     END;
   PROCEDURE MULTIPLEOSC;
     BEGIN
            SETANGLE(-45);
            OSCILLATE;
            CALCDATASPACE;
            DELAY(315);
            SETANGLE(45);
            OSCILLATE;
            CALCDATASPACE;
            DELAY(315);

     END;
   PROCEDURE OSC;
      BEGIN
            N:=0;
            WHILE N<12 DO
            BEGIN
             MULTIPLEOSC;
             N:=N+1;
             END;
             ID;
             WRIT;
              INITSPACE;
       END;
  BEGIN
      WRITELN;
      WRITELN;
      WRITELN;
      WRITELN('ENTER THE DATE');
      READ(DATE);
          I:=0;
          N:=0;
          K:=MEM[STOREST]+MEM[STOREST+1]*256-16;
          ID;
   MENU:
       WRITELN('SET A/D CHANNEL 0                    0');
       WRITELN('SET A/D CHANNEL 1                    1');
       WRITELN('START OSCILLATING ONCE               O');
       WRITELN('START SERIES OF OSCILLATIONS         S');
       READ(S);
       CASE S OF
       '0': SETTRIG0;
       '1': SETTRIG1;
       'O': BEGIN
```

```
                        SETANGLE(0);
                        OSCILLATE;
             END;
     'S':    BEGIN
              REPEAT
              BEGIN
               OSC;
               I:=I+1;
               WRITELN(I);
              END;
              UNTIL I>120;
              END;
              END;
   I:=0;
   GOTO MENU;
   END.
```

CHAPTER 2

Computer analysis of electrophysiological signals

JOHN DEMPSTER

1. INTRODUCTION

Electrophysiologists have always been attracted to the use of computers for enhancing and speeding up the analysis of their experimental results. Electrophysiological methods have, since the late 1940s, produced much valuable information concerning the underlying mechanism of electrical excitability in nerve and muscle cells. In particular the experimental control of the cell membrane potential using the 'voltage-clamp' method has allowed the dissection and study of a wide variety of trans-membrane ion (Na^+, K^+, Ca^{2+}, Cl^-) conductances. More recently, the 'patch clamp' method (see Section 5.10), in addition to visualizing the flow of current through a single ion channel in the cell membrane, has allowed the electrophysiological technique to be extended to a much wider range of cell types, especially in central nervous system. Before the use of computers, electrophysiological signals, due to their short duration, could only be captured on an oscilloscope screen and recorded on 35 mm film. The film then had to be developed and the enlarged record measured manually. This process was time consuming and limited the achievable measurement resolution. The introduction by Digital Equipment Corporation of the minicomputer in the 1970s, first the PDP8 and then the PDP11, allowed the computer to be used in the electrophysiological laboratory for the first time. Nevertheless, minicomputers were still high cost items and only a relatively small number of laboratories could afford one. The first generation of microcomputers (PET, Apple II) while finding many other useful laboratory applications, were insufficiently powerful to replace the minicomputer.

The situation has now drastically changed with the arrival of the second generation of microcomputers, typified by the IBM personal computer (PC). In most respects they are equal, if not superior, to the minicomputers of the 1970s, at a fraction of their cost. Certain features of the IBM PC family of computers make them particularly suitable for electrophysiological work. They are widely available in a variety of cost versus performance configurations. They have expansion slots allowing easy installation of laboratory interface cards and an optional maths co-processor which greatly enhances arithmetic computation speed. A wide range of available software also makes them useful for word processing, statistical data analysis, or storing references.

This chapter discusses the application of the microcomputer to the analysis of electrophysiological signals.

Figure 1 shows a diagram of a computer system, which could be used to collect and analyse electrophysiological signals. It is based around an IBM PC AT (or compatible)

Computer analysis of electrophysiological signals

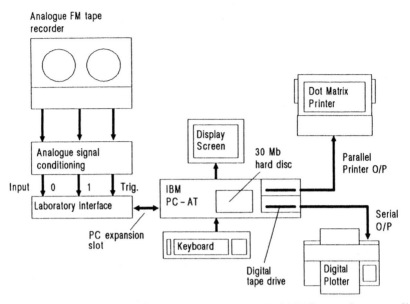

Figure 1. An electrophysiological data analysis system based on an IBM PC AT personal computer. Signals recorded on FM tape are played back, digitized using a laboratory interface unit, and stored on disc. Results are displayed on screen or plotted on a dot matrix printer or digital plotter.

personal computer. Signals recorded on an FM analogue tape recorder are played, via signal conditioning amplifiers, into a laboratory interface unit which converts the signals into digital form. The digitized signals are stored on computer disc and can be displayed on a screen, printer or plotter. A detailed description of the hardware comprising this system follows in Sections 2–4, while software for electrophysiological analysis is discussed in Section 5. Section 6 discusses the development of electrophysiological software.

2. COMPUTER HARDWARE FOR DIGITIZING ANALOGUE SIGNALS

2.1 The laboratory interface unit

Most computer systems do not include any means of recording signals from laboratory experiments as a standard feature. They require a laboratory interface installed in an expansion slot, which allows the computer to record or manipulate the variety of electrical signals found in the laboratory. An interface consists of an analogue-to-digital converter (ADC) for converting a continuously varying (analogue) voltage signal into binary digital form, a programmable clock capable of timing the intervals between ADC samples, a digital-to-analogue converter (DAC) for producing an analogue voltage under computer control and digital input and output lines allowing the computer to detect and control the state of switches, valves, or other processes related to an experiment. The more powerful laboratory interfaces have on-board microprocessors with their own RAM storage memory and operating programs, making them small special purpose computers in their own right. These intelligent interfaces greatly improve the overall performance of the system.

Table 1. Specifications of some laboratory interfaces in common use in electrophysiological laboratories.

	CED 1401	DT2801A	DT2821	DT2828	Tecmar Labmaster
A/D conversion					
No. of channels	16	16	16	4	16
Resolution	12	12	12	12	12
Max. sample rate (kHz)	80 (150[a])	27.5	50 (130)[a]	100	30 (80)[a]
D/A conversion					
No. of channels	4	2	2	2	2
Resolution	12	12	12	12	12
Max. rate (kHz)	80 (150)[a]	27.5	50 (130)[a]	100	200
Digital input/output (I/O) lines	16	16	16	16	24
DMA supported	YES	YES	YES	YES	NO
Computers	AT/PC	AT/PC	AT	AT	AT/PC
Special features[b]	1,5	2	2,3	3,4	2,5

[a]Brackets indicate available as special option.
[b]The special features indicated are as follows: (1) User written programs can be down-loaded from PC into interface, capable of simultaneous A/D and D/A conversions to/from on-board data buffers. Optional 8 Mbyte RAM storage; (2) Programmable ADC input voltage with ×1, 2, 4, 8 amplification; (3) Two linked DMA channels available to support continuous ADC sampling to disc; (4) ADC sampling simultaneous on all four channels and (5) Five on-board programmable counter timers.

2.1.1 *Analogue-to-digital conversion*

Although all ADCs perform the same basic operation, the speed and the resolution of conversion may vary. The resolution of an ADC refers to the number of binary digits (bits) available to represent the digitized analogue signal, the most common values being 8, 12 or 16 bits, splitting the analogue voltage range into 256, 4096 and 65536 units, respectively. In practice, at least 12-bit resolution is required to ensure an adequate digitized representation. On the other hand, the noise levels of most electrophysiological recording systems are not sufficiently low to make use of 16-bit resolution. For this reason most laboratories currently use 12-bit ADCs (see Chapter 1).

Many of the same criteria as discussed above for ADC conversion (resolution, conversion rate and output voltage range, etc.) apply equally to digital-to-analogue (D/A) conversion. *Table 1* contains the specifications of some of the more popular laboratory interfaces currently in use in electrophysiological laboratories.

2.1.2 *Digital sampling of analogue signals*

Since the process of analogue-to-digital (A/D) conversion takes a short but finite period of time, an analogue signal can only be sampled at discrete intervals, with the minimum interval determined by the time taken to perform an A/D conversion. It is not, of course, always necessary to sample at the maximum possible rate. A rate between five and ten times the highest frequency component in the signal under study will suffice for most purposes. This may vary between rates of 1 Hz where slow signals, such as changes in cell resting potential, are being studied to rates of 25−100 kHz when studying very brief signals, such as neuromuscular junction end-plate currents or nerve sodium currents which last less than 2−3 msec. The time interval between ADC samples must be strictly controlled, usually by a programmable clock within the laboratory interface. Such clocks produce pulses at multiples of some fixed rate, determined by the value preset into the clock's counter. When the clock is started it sends a stream of digital pulses to the ADC at these fixed intervals, each pulse initiating an A/D conversion. Usually an option also exists to synchronize the start of the clock pulse train with an external trigger signal.

2.1.3 *Multi-channel sampling*

Laboratory interfaces generally have more than one ADC input channel, 8 or 16 channels being common, with a multiplexer (a fast digitally controlled switch) being used to sequentially connect each input to the A/D converter which then performs the conversion. The effective maximum sampling rate is reduced to the single channel maximum rate divided by the number of channels. To ensure a high performance when sampling more than one channel, multiplexing is controlled by the on-board microprocessor which automatically sequences through the chosen set of channels. Since channels are sampled one at a time, multi-channel samples are inevitably skewed by one sampling interval between each channel in the collection sequence. If groups of channels must be sampled at exactly the same time an interface which supports simultaneous multi-channel sampling must be used, such as the Data Translation DT2828. This interface has only four input channels, but an analogue sample and hold (S/H) circuit for each channel and a very fast ADC. The analogue signal on each channel is frozen at exactly the same instant and held by the S/H circuit, for as long as is required by the converter to sequence through and digitize each channel.

2.1.4 *Controlling the laboratory interface from software*

Programs running on the PC communicate with a laboratory interface through hardware I/O ports (or registers). Each I/O port on the interface card is assigned a unique port address number and data can be written to it, or read from it, using the OUT and IN 8086 machine code instructions. In general, interfaces from different manufacturers differ significantly in the number and function of I/O ports. Port addresses also vary, but can often be set using switches on the interface circuit board to avoid conflicts with other devices. Most interfaces, however, have at least one port which acts as a control/status (C/S) register, showing the current status of the interface when read and accepting interface operating commands when written. There is usually also a data register which is used to transfer ADC or DAC data to or from the interface. Since controlling an interface by means of its I/O ports normally requires an understanding of machine code programming, most manufacturers provide a software subroutine library to simplify operation. The BASIC program listings in *Figure 2* show how two different laboratory interfaces, the Data Translation DT2801A and the Cambridge Electronic Design 1401 perform the task of collecting a single ADC sample. In the DT2801A program (*Figure 2a*) the interface is being handled by directly manipulating its I/O ports using the BASIC instructions, OUT, INP() and WAIT. This is contrasted with the 1401 program (*Figure 2b*) which uses a software device driver allowing the interface

(a)

```
100                          ' Get ADC sample from DT2801A
110 CSR = &H2ED              ' Control/Status I/O port
120 DR = &H2EC               ' Data I/O port
130 WAIT CSR,2,2             ' Wait till data port empty
140 WAIT CSR,4               ' Wait till DT2801A ready for command
150 OUT CSR,&H0C             ' Request ADC conversion
160 WAIT CSR,2,2             ' Wait till data port empty
170 OUT DR,1                 ' Select X2 ADC input gain
180 WAIT CSR,2,2             ' Wait till data port empty
190 OUT DR,0                 ' Select ADC channel 0
200 WAIT CSR,1               ' Wait for byte to appear in data port
210 LOW.BYTE = INP(DR)       ' Read 1st byte from data port
220 WAIT CSR,1               ' Wait for byte to appear in data port
130 HIGH.BYTE = INP(DR)      ' Read 2nd byte from data port
140 ADC.VALUE = HIGH.BYTE*256 + LOW.BYTE ' Combine into 12 bit no.
```

(b)

```
100                                  ' Get ADC sample from 1401
110 OPEN "LAB0" FOR OUTPUT AS #1     ' Open output file stream to 1401
120 OPEN "LABI" FOR INPUT AS #2      ' Open input file stream
130 PRINT #1, "ADC,0;";              ' Request ADC sample from ch.0
140 GET #2                           ' Wait for response from 1401
150 INPUT #2, ADC.VALUE              ' Collect ADC sample
```

Figure 2. BASIC program code for collecting a single ADC sample from a laboratory interface. (a) For a Data Translation DT2801A interface being controlled by writing directly to the I/O ports of the interfaces. (b) For a Cambridge Electronic Design 1401 with a software driver allowing the interface to be controlled by means of high level commands.

to be operated by means of a set of high level commands without any need to be aware of the I/O port addresses.

2.2 Data transfer from interface to computer

2.2.1 Programmed data transfer

When A/D sampling is in progress a 12-bit binary number is produced after each conversion in the A/D converter data register. It must be removed before the next conversion can start and stored in memory. There must therefore be space set aside either in PC memory, or in memory within the interface, for these samples. The simplest way to collect a series of ADC samples is to write a program to monitor the contents of the status register and wait until the 'conversion done' flag becomes set and then to read the latest sample from the data register and copy it into the memory buffer. While this method, known as 'programmed I/O', is fast and suitable for collecting short sequences of a few thousand samples it has a number of disadvantages due to the fact that the central processor unit (CPU) is completely committed to repeatedly checking the contents of the status register. Sampling rates are limited to around 30 kHz (70 kHz for AT computers) by the time taken to execute the status testing/sample storage loop. In addition, it is possible to completely miss the occasional group of samples if there are any delays in the program loop such as introduced by time taken out for the CPU to service keyboard or real time clock interrupts. Interrupts must therefore be turned off (using the CLI assembler instruction) while sampling is taking place.

2.2.2 *Interrupt driven data transfer*

Normally when a program is running CPU instructions are executed one after the other in a fixed sequence. In order to allow the program to escape from this sequence and respond to events which occur asynchronously, a hardware feature known as the 'interrupt' can be used. The PC expansion bus has eight interrupt request lines (IRQ0...IRQ7) which allow a peripheral device installed in the expansion bus to interrupt the currently executing program and to transfer control to a small subroutine known as an interrupt service routine. If one of these lines can be activated by the laboratory interface when an A/D conversion is done it can be used to invoke a routine to transfer the sample into memory, removing the need for the data collection program to constantly test the status register. In addition, each time an interrupt is requested, the current state of the CPU data registers and position in the program must be saved before the interrupt service routine is executed. The overheads involved in these processes limit sampling rates when using interrupts to around 10 kHz. Although slower than programmed data transfer, interrupt driven transfer is more robust since it ensures each sample is automatically transferred as it is produced by the interface. In addition, between interrupts the main program can perform other tasks such as writing previously collected data to a disc file.

2.2.3 *Direct memory access data transfer*

The overheads associated with interrupt-driven data transfer can be avoided by using a direct memory access (DMA) controller to transfer data between the laboratory interface and the PC memory. This device (an Intel 8237A in the PC) is capable of transferring data to and from the laboratory interface on request, without requiring the services of the 8086 CPU. When the interface requests a data transfer (by activating a DMA request line on the PC expansion bus), the 8237A locks out the CPU, takes control of the computer bus, completes the transfer, and returns control to the CPU. Since the CPU is not involved in the data transfer there is no overhead in saving the CPU data registers. A single DMA transfer channel is capable of transferring data at rates up to 750 Kbytes/sec. This method therefore allows very high sampling rates limited only by the ADC conversion speed of the interface. Before a DMA data transfer can take place the direction of data transfer, starting address of the data buffer, number of bytes to be transferred and a number of other initialization details must be programmed into the DMA controller. The 8237A DMA controller on the PC supports four separate DMA transfer channels with one of them free for use by a laboratory interface (AT computers have six spare channels). Appropriate software is often provided by the supplier of the laboratory interface but, if not, it is possible to program the controller I/O ports directly (3). DMA is, by far, the fastest and most versatile means of transferring data between interface and memory and is essential to support the software described in Section 5.

3. SIGNAL CONDITIONING

3.1 Analogue signal conditioning

The analogue inputs of the laboratory interface have either a fixed input voltage range (usually +/− 5 V) or a limited range of programmable options (e.g. +/− 10, 5, 2.5, 1.25 V for the Data Translation DT2801A board). If the available analogue signal spans only a small proportion of the voltage range then it will not be possible to take full advantage of the ADC resolution. In such cases it is useful to be able to amplify the signal so that it spans approximately 75% of the range. The gain of this amplifier should be variable using either a multi-turn potentiometer and calibrated dial or a switchable series of fixed gains. Signals which are predominantly positive (or negative) going relative to their baseline can only make use of half of the analogue input voltage range, effectively losing 1 bit of ADC resolution. It is therefore also useful to be able to shift the signal baseline, before amplification, to allow it to span a greater proportion of the voltage range.

3.2 Filtering signals

3.2.1 *Low-pass anti-aliasing filters*

Periodic waveforms such as sine waves cannot be adequately represented digitally

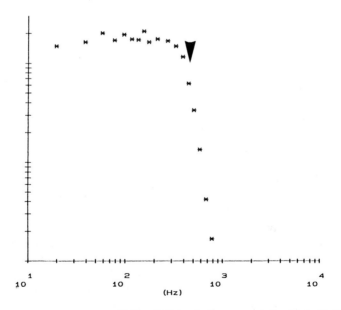

Figure 3. Frequency response spectrum (10 Hz − 10 kHz) of a low-pass anti-alias filter with 400 Hz cut-off frequency (indicated by arrow). Filter is a 4-pole Butterworth design with a 24 dB per octave roll-off after the cut-off frequency (implemented using a MF10-CM switched capacitor filter IC).

without at least two sample points per cycle. In addition, the digital record of a waveform which has been 'under-sampled' appears to contain frequency components quite different from those in the original analogue signal, a phenomenon known as 'aliasing'. Nyquist's theorem states that a signal must be sampled at a rate at least twice that of the highest frequency component within the signal in order to prevent aliasing. In practice, however, to achieve a visually pleasing signal record one would sample at a rate substantially higher than this ($5-10\times$). To avoid aliasing problems analogue signals are passed through a low-pass filter to ensure that all frequency components above the Nyquist frequency are removed before digitization. Although low-pass filtering can be achieved using passive components such as resistors and capacitors, better performance can be achieved using active filters constructed using operational amplifiers with networks of resistors and capacitors in their feedback loop. Such filters are available either as integrated circuit building blocks which can be incorporated in one's own designs or as complete self-contained filter units (Frequency Devices). The performance of a filter is described by its frequency response spectrum, an example of which is shown in *Figure 3*. The pass-band of the filter is the region from DC up to the cut-off frequency (f_c) defined as the frequency at which the filter output power will be attenuated by 50%. Filter attenuation factors are commonly described in units of decibels (dB) which are logarithmic units of the ratio of the input and output signal power

$$1 \text{ dB} = 10 \times \log_{10}(P_{out}/P_{in})$$

At f_c the signal attenuation is therefore -3 dB. The rate at which the filter attenuates the signal with increasing frequency after f_c is described as the roll-off in units of dB per octave ($\times 2$ increase in frequency) or decade ($\times 10$ increase).

3.2.2 Bessel, Chebyshev and Butterworth filters

An ideal filter would be one which entirely removed all frequencies above the cut-off and passed all the rest without distortion. Unfortunately, such an ideal frequency response cannot be achieved in actual filters. There are, however, a variety of filter circuit designs which achieve one or other of these criteria. Chebyshev and Butterworth filters have good roll-off performances and rapidly remove frequencies above f_c, but distort the signal waveform by introducing substantial ($10-20\%$) overshoot and ringing into step-like waveforms. On the other hand, Bessel filters have a shallow roll-off but do not significantly distort the signal. The roll-off figure for a filter design can be improved by feeding the signal through a series of identical filter subunits. The number of sub-units in a filter is expressed in terms of 'poles' (factors in the denominator of the filter transfer function) with each subunit normally contributing two poles. Commercially available filters such as the Frequency Devices 902 consist of four subunits and therefore are 8-pole filters. Filter manufacturers generally provide both Bessel and Chebyshev or Butterworth versions of their filters. A useful discussion of the relative performance of different types of filters can be found in Chapter 4 of reference 4. For most electrophysiological work it is more important to achieve minimum distortion of the filtered signal than to have a rapid roll-off. Therefore Bessel filter designs should be chosen in preference to Chebyshev or Butterworth designs. However, power spectral analysis (see Section 5.8) is an exception. It is very important to ensure that frequency components

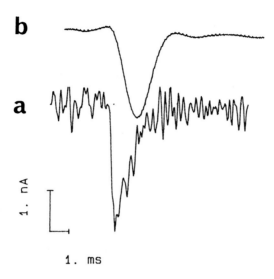

Figure 4. Effects of excessive low-pass filtering on the time course of a signal. (**a**) Digitized record of a miniature end-plate current filtered at the normal 5 kHz cut-off frequency. (**b**) Same signal filtered with 400 kHz cut-off.

above the Nyquist frequency (which corresponds to the upper limit of the power spectrum) are removed as effectively as possible to prevent them appearing as aliased frequencies within the spectrum causing considerable distortion. A Chebyshev or Butterworth filter should therefore be used when digitizing signals for spectral analysis.

3.2.3 *Other effects of low-pass filtering*

When choosing the filter cut-off frequency consideration must be given to any effects that it might have on the signal of interest. In addition to preventing aliasing, low-pass filtering cuts out the high frequency components of the signal background noise, improving the signal-to-noise-ratio. However, the filter will also distort the waveform of the signal itself if any of it's own components extend beyond the filter pass-band. *Figure 4* illustrates the effect of the filter f_c on a digitized record of a miniature end-plate current (MEPC), a signal with a fast time course (and hence high frequency components) and a high background noise. *Figure 4a* shows the MEPC as it would normally be recorded, low-pass filtered with a cut-off frequency of 5 kHz. *Figure 4b* shows the same MEPC filtered at 400 Hz. It is apparent that the peak signal-to-noise-ratio has been markedly improved by reducing the cut-off frequency [2.5:1 in (a) compared to 11:1 in (b)]. However, the signal waveform has also been grossly distorted with both the rise and decay times being prolonged and the size of the MEPC peak significantly reduced. The optimal choice of f_c is therefore one which achieves as much improvement in signal-to-noise-ratio as possible without distorting the signal waveform. In practice, to find the optimum it is useful to digitize the signal using a range of different cut-off frequencies starting at a high value then decreasing until distortion of the signal is observed.

Computer analysis of electrophysiological signals

3.3 Synchronization trigger signal

Laboratory interfaces generally have an external input which can be used to initiate a train of ADC samples. This is usually a digital, active-low, TTL-compatible input. TTL (transistor−transistor−logic) is a type of digital circuitry in which 0 V ($\leq$ 0.4 V to be precise) is defined as the 'LOW' or 'OFF' state and 5 V as the 'HIGH' or 'ON' state. 'Active-LOW' signifies that it is the LOW state which acts as the trigger. Unfortunately, most synchronization signals produced by stimulators or recorded on FM magnetic tape are incompatible with this kind of TTL input, providing pulses from 0 V to some positive or negative value rather than the required 5 V to 0 V pulse. Signal conditioning is therefore required to hold the input at the 5 V level and pulse it to zero when the synchronization signal is detected.

3.4 A simple signal conditioning circuit

Figure 5 shows the circuit diagram of a signal conditioning circuit used to amplify a 1 V peak signal from a channel of an FM tape recorder to match the 5 V range of a laboratory interface. It provides four switched gains (5, 10, 15, 25$\times$) and a DC offset facility. Conversion of the 1 V synchronization pulse (recorded on a second tape channel) is performed using an LM311 comparator with a variable threshold level (circuit adapted from p. 389 of ref. 4). While it is possible to construct a signal conditioning

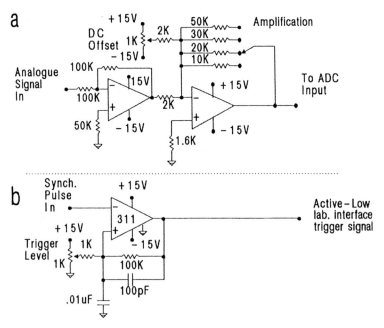

Figure 5. An analogue signal conditioning circuit designed to match the 1 V peak output of an FM tape recorder to the 5 V input range of a laboratory interface. Circuit (**a**) provides 5,10,15,25$\times$ gain and +/− 15 V DC offset. Operational amplifiers are OP-07s, resistors should be 1% precision. Circuit (**b**) is a 311 comparator providing a variable threshold trigger circuit for the 'active-low' TTL trigger input of the laboratory interface.

system from easily available operational amplifier circuit components, complete signal conditioning packages are also available commercially. The Neurolog system of rack mounted signal conditioning modules is a typical example (Digitimer Ltd). A two channel (voltage and current) system would comprise two differential DC amplifiers with calibrated variable gain ($0-10\times$ or $0-100\times$) and DC offset (NL106), each with a high/low pass filter in series (NL125), and a threshold detector used to produce the TTL trigger synchronization signal (NL200).

4. COMPUTER HARDWARE

4.1 The IBM PC and compatible computers

IBM PC-compatible computers are based on the 80×86 family of 16- or 32-bit microprocessors, the 8088, 8086, 80286, 80386 listed in order of increasing performance. The orginal IBM PC, first marketed in 1981, was based on an 8088 with a microprocessor clock speed of 4.47 MHz and had 64 Kbytes of memory. While this particular machine must now be considered to be obsolete, the central aspects of its design has become a *de facto* world standard. An IBM-compatible computer must be able to run all of the software written for the true IBM PCs and it must be able to accept expansion cards designed for the PC expansion bus. Most large computer companies (and many small ones) now provide at least one IBM-compatible PC as part of their product range. These machines although adhering to the IBM standard may nevertheless vary considerably in price, performance and styling. Currently available PCs are equipped with at least an 8 MHz 8088 or 8086 microprocessor and 512 Kbytes of memory and are two to three times faster than the original IBM PC. PCs based on the 80286 (IBM AT) and the 80386 (Compaq Deskpro 386) with clock speeds of $8-20$ MHz however are substantially faster than these 8088/8086 machines by as much as $2-5\times$. This is due, not only to the increased clock speed and hence faster execution of instructions, but also to improvements in the internal design of the microprocessor. They also can have as much as $4-8$ Mbytes of memory installed compared to the upper limit of 640 Kbytes for the 8088/8086. The improved performance of the 80286/80386 class of PCs make them particularly suitable for the computation-intensive tasks found in the analysis of electrophysiological signals. The optional maths co-processor (8087, 80287, or 80387) should always be included in a PC intended for scientific applications since it provides a 20 times improvement in the speed of floating point calculation. With a co-processor installed, the time taken for calculations such as iterative curve fitting (see Section 5.4) can be reduced from 10 min to 30 sec.

In 1987 IBM introduced a successor to the IBM PC range of computers, the Personal System 2 (PS/2) capable of running existing PC software but offering, higher resolution graphics in more colours, higher performance in some respects, and greater potential for future expansion. Unfortunately, at the time of writing (July 1987) no laboratory interface is available for the new micro channel architecture expansion bus of the PS/2s. It is likely however that it will not be long before this omission is remedied.

4.2 The MS-DOS operating system

All IBM-compatible computers make use of versions of the MS-DOS disc operating system (called PC-DOS for genuine IBM PCs) produced by Microsoft. This operating system has evolved from version 1 on the original IBM PC to the current version 3.3 on the latest IBM machines. Almost all of the programs that have been developed for the PC require the use of MS-DOS. Although other operating systems exist, notably the multi-user XENIX (a version of UNIX), they have failed to gain general acceptance and are found only in specialist areas. MS-DOS is a single-user, single-task operating system, designed to support a single screen and keyboard and to run one program at a time. This is not a great disadvantage for the kind of signal collection and analysis described here and, in fact, the simplicity of MS-DOS, compared to a a multi-user, multi-task system such as XENIX, makes software development much easier.

4.3 Data storage

4.3.1 *Magnetic storage discs*

The sampling of electrophysiological signals at a high rate rapidly generates more digitized data than can be stored within the computer's memory. When this happens, and also for permanent storage, the data must be transferred to a magnetic disc of some form. PCs are usually equipped with at least one 5.25 inch floppy disc drive. IBM PC and PC XT class computers use a double-sided 360 Kbyte capacity disc while PC AT class computers use a high density 1.2 Mbyte capacity disc. On some machines, particularly the IBM AT itself, these two types of disc drive are only partially compatible. Data written on to a 360 Kbyte disc can be reliably read by a 1.2 Mbyte drive. However, 360 Kbyte discs written using a 1.2 Mbyte drive cannot always be read reliably by a 360 Kbyte drive, due to differences in the width of the data track written by each type of disc drive.

Floppy discs, although cheap and removable, have a small storage capacity and are slow compared to the other common type of disc storage, the Winchester hard disc. Winchester discs are rigid non-removable magnetic storage discs, contained in a hermetically sealed case to exclude dust particles from the disc surface, and rotating at much higher speed than floppy discs. Currently, storage capacities range from 10 to 150 Mbytes. The speed of reading and writing to the disc is especially important when there is a requirement to collect and store on disc long unbroken sequences of samples at a high sampling rate. The crucial factor is the time taken to move the disc read/write head radially from track to track. In the disc specifications this is quantified by the average time taken to access a particular sector on the disc (average access time). The speed of the hard disc installed in computers by different manufacturers can vary considerably with average access time from 15 to 80 msec. In general, the PC XT and almost all similar compatibles have a slow disc with access times in the range 60–80 msec while the IBM AT has a disc with an access time of 33 msec. AT-compatibles from other manufacturers, however, do not always have similarly fast discs, and it is worth enquiring about this when purchasing a PC.

4.3.2 *Long-term data storage*

Some form of removable storage medium is required since even apparently large capacities of 30 Mbytes or more cannot store more than a few weeks of experimental results. If individual data files are sufficiently small then they can be copied on to floppy discs. A 1.2 Mbyte disc, for instance, can hold a single file containing over 1000 signal records of 512 ADC samples each. If files are larger than 1.2 Mbytes then either a digital tape drive or a high capacity removable disc is required. Digital tape drives have storage capacities comparable to hard discs (10−40 Mbytes per tape), are relatively inexpensive, but are very slow with transfer rates of 1−5 Mbytes per min. Tape drives are available using a variety of tape cartridges (DC300, DC2000) and even video tape. There is currently no single standard tape format as there is for floppy discs and therefore it is unlikely that data stored on a tape cartridge by one tape drive can be read by a different manufacturer's drive. Similarly, files are transferred to the tape drive using special utility software provided with the tape drive since standard MS-DOS file structures are not used.

A small number of high capacity removable disc systems are also available and provide a faster but more expensive alternative to tape. Best known among these is the 'Bernouli box' disc cartridge with capacities ranging from 10 to 80 Mbytes. High capacity discs and drives are difficult to manufacture since the discs must rotate at high speed and the read/write heads must be positioned very close to the disc surface. The discs must therefore be kept scrupulously free of dust particles large enough to collide with the heads. An interesting solution to this problem is the Tandon Data-Pac which consists of a complete sealed 30 Mbyte Winchester disc plus read/write head assembly, which can be removed as a unit from the rest of the drive.

4.4 Display of graphics

Most PCs now have the capability to draw graphical information on the screen but there is a wide variety of display adapters supporting different screen resolutions and ranges of colours. The oldest and most basic display is the IBM colour graphics adapter (CGA) which can provide a maximum resolution of 640 h × 200 v pixels (picture elements = dots) in two colours. It also supports four colours with a reduced resolution of 320 h × 200 v. In practice, 640 h × 200 v is the minimum usable, allowing the display of a 512 point signal record with approximately 0.5% accuracy. It is worth noting here that for inspection purposes there is no need to provide a display resolution which matches that of the data (e.g. 4096 h × 4096 v for 12-bit ADC samples). As long as the signal record is stored somewhere with its full resolution it is always possible to zoom in on detailed portions of the signal record before display.

Many other manufacturers have capitalized on the limitations of the CGA by providing alternative display adapters with better resolution or range of colours. One particularly successful one, the Hercules monochrome display adapter, provides a significantly higher resolution (720 h × 350 v) and has become a display standard in its own right. This display provides a significantly better rendition of alphanumeric characters since they are composed of a 9 × 14 pixel matrix rather than the 8 × 8 available on the CGA. IBM introduced its enhanced graphics adapter (EGA) in 1984 which provides a resolution of 640 h × 350 v in 16 colours. The development of display adapters has not ceased with IBM producing a new range of adapters (MCGA, VGA) for its Personal System/2 range of computers. These existing display adapters rely on the PC to do all of computations related to generating lines, curves and the movement of blocks of pixels on the display screen. As the display resolution increases this becomes a slower and slower process. A new generation of display adapters are currently becoming available with their own on-board display co-processor, such as the Texas Instruments TMS34010 or the Intel 82786. These adapters will provide a $100\times$ increase in the speed of drawing on the screen due to the co-processor taking over many of the tasks of display generation.

4.5 Hard copy

4.5.1 *Dot matrix printers*

When a hard copy of data displayed on the computer screen is required a variety of devices can be used, but the dot matrix printer is probably the least expensive and most versatile. It is fast, with printing speeds of 100−300 characters per second (CPS), and capable of printing both text and graphics as required. Characters are formed on the page by the impact of a group of nine moving print wires pressing through an inked ribbon, as the print head is swept across the paper. Such printers have three distinct modes of operation. In the 'draft' mode the printer simply prints characters from one of a small variety of in-built typefaces or styles such as bold, underlined or italic. This is the fastest printing mode but characters have a poor resolution consisting of a 5 h × 7 v matrix of points within a 9 h × 9 v field. Many printers however have a slower 'Near Letter Quality' (NLQ) mode in which a higher character resolution is achieved by using two printing passes over each line. The print quality in this mode is often acceptable for correspondence or producing manuscripts.

Graphs and diagrams can also be plotted on a dot matrix printer by using the 'bit image' mode. In this mode each pin in the printer is directly controllable from the PC allowing images to be built up on the page by sweeping the print head across the paper in a raster fashion. The most commonly used image transfer protocol for this was developed by Epson and is a standard feature on all their printers and many others, including those produced by IBM. In bit image mode, as for NLQ mode, using two passes of the print head per line allows a reasonably good resolution of 120 dots per inch (DPI) equivalent to a pixel resolution of 1020 h × 1440 v for an A4 page, to be produced in 1−2 min.

Since the image on the display screen is also a dot pattern it possible to transfer it line by line to the printer. MS-DOS, in fact contains a program to do this (GRAPHICS.COM) which 'dumps' the screen display to the printer when the Print Screen key is pressed. While this is the simplest way of passing graphics to the printer it has the major disadvantage of limiting the graphics resolution to that of the display. Images produced in this way therefore are rarely suitable for publication. If one wishes to make use of the full printer resolution, a high resolution bit image must be created in the computer memory then passed to the printer. This is rather a complex task since there is no simple way of drawing lines, symbols, or even text characters in bit-image mode. It is however possible to obtain graphics software packages such as GEM (Graphics Environment Manager, Digital Research) which contain the required software and make the task of plotting high quality graphs on dot matrix printers feasible (see Section 6.4). Compare *Figure 7*, for instance, from a screen dump with *Figure 9* produced using GEM.

Table 2. Specifications of common graphics display adapters for IBM PC.

Adapter	Resolution	Colours	Computer
Colour graphics adapter (CGA)	640 h × 200 v	2	PC/AT
Enhanced colour graphics (EGA)	640 h × 350 v	16	PC/AT
Professional graphics (PGA)	640 h × 400 v	256	PC/AT
Hercules	720 h × 350 v	2	PC/AT
Hercules In-color	720 h × 350 v	16	PC/AT
Video graphics array (VGA)	640 h × 480 v	16	PS/2
IBM graphics printer	1020 h × 990 v	2	-
Hewlett Packard HP7475	10900 h × 7650 v	6	-
Apple Laserwriter	2475 h × 3600 v	2	-

4.5.2 Digital graph plotters

Higher resolution and better formed lines can be achieved using a digital graph plotter which draws lines or text on plain paper using a fine fibre tipped pen. Plotters are usually connected to the PC by means of the serial (RS232) interface. The plotter receives instructions from the PC in the form of a graphics language which specifies the moves that the pen has to make to draw a graph. The Hewlett Packard range of plotters make use of a code called the Hewlett Packard Graphics Language (HPGL) which is also supported by other manufacturers. Plotting graphs on such plotters is much easier than on dot matrix printers since HPGL provides instructions for the drawing of lines and text in various styles and colours. Hewlett Packard produce a series of digital plotters from the low-cost HP7440 (8 pen, A4), HP7475 (6 pen, A3) to the HP7550 (8 pen, A4 with sheet feeder). HPGL-compatible plotters include the Gould 6120. *Figure 8* was produced using an HP7440 plotter.

4.5.3 Laser printers

Laser printers are essentially computer controlled photocopiers. Instead of the optical system of a normal photocopier, the image is written by a laser beam on to the electrically charged photo-sensitive drum which then picks up toner and transfers it to paper. Laser printers are like dot matrix printers in that they are fast and have text and bit image modes. Printed characters are however of a much higher quality and graphics resolutions of 300 × 300 DPI are common. In addition some laser printers such as the Apple Laserwriter have sophisticated graphics codes such as Postscript, capable of drawing lines and text in various sizes, styles, and orientations. Although, laser printers can effectively combine the roles of the dot matrix printer and the plotter their higher cost currently limits the use of laser printers in the laboratory. *Table 2* lists the specifications of some graphics displays and plotting devices.

5. A SOFTWARE PACKAGE FOR ANALYSING ELECTROPHYSIOLOGICAL DATA

5.1 Electrophysiological analysis software

The software discussed here is a set of analysis programs developed over a period of years for use by the Neuromuscular Research Group at the University of Strathclyde and by groups working in similar fields, particularly at the Universities of Dundee and St Andrews. The experiments performed in these laboratories utilize a wide range of techniques, including two microelectrode voltage-clamp recording, patch clamp recording (single channel and whole cell), and noise analysis. The software therefore covers a similar range of analysis procedures. Although the details of the analyses vary from one type of experiment to another, certain general categories of operation can be found common to all the programs.

(i)   Digitization of analogue signals.
(ii) Display and validation of digitized signals.
(iii) Enhancement of stored signals (e.g. averaging).
(iv) Quantitative measurement of signals (e.g. peak amplitudes).

Each of these operations is incorporated into its own discrete software module, largely independent of the others, which can be chosen by the user from a master menu displayed on screen. Unfortunately, since the programs to be described are somewhat large, the source code cannot be presented here. This section, therefore, presents an overview of the software structure, typical data screen displays and results, and a discussion of key performance issues.

5.1.1 *Analogue signal digitization*

The digitization module performs the transfer of a stream of ADC samples from the laboratory interface into RAM within the PC and then onto disc, creating a data file containing a digitized copy of all of the signals to be analysed. The data stored on this file can then be accessed by the other display and analysis modules. This is usually a more practical method than to attempt to digitize signals in a piece-meal fashion as they are required for analysis. To minimize access times and demands on disc storage space, ADC samples are stored in this file in their raw binary form. One ADC sample in binary form takes up 2 bytes of storage while, if converted to the readable ASCII code, at least 4 bytes are required. Samples are also read and written to files in multiples of 512-byte blocks (disc sector size) since disc transfers are more efficent in these sizes. Key parts of the digitization module are written in assembler language since it must execute as fast as possible. Also many of the operations involved in operating the analogue-to-digital converter such as the reading and writing to I/O ports are more conveniently done in assembler language than in higher level languages such as FORTRAN.

5.1.2 *Continuous signal records*

There are two main approaches to digitizing analogue signals. One is continuous collection where a single long record, containing an unbroken series of samples, is collected. This approach has the advantage that one need not make any attempt to detect where the signals lie within the record at the time of collection. It is particularly suitable in cases where the events of interest occur frequently but at random intervals such as the fluctuations in the current through single ion channels. A critical factor in continuous collection is the highest sampling rate and maximum size of record that can be collected at that rate.

If the whole record can be stored within RAM then very high samplings rates (150 kHz) can be achieved. Interfaces which use a storage buffer in PC memory are often limited in size to as little as 64 Kbytes, by having to share memory with the program code. At high sampling rates, this can drastically limit the total length of record collected since at a rate of 100 kHz this buffer will be filled after only 320 msec. Although some interfaces can have large amounts of on-board RAM installed (8 Mbytes on CED1401) this is still substantially less than can be stored on disc (32 Mbytes).

If slightly lower samplings rates are acceptable then very much longer records can be collected by transferring blocks of ADC samples to a disc file as soon as they are collected, requiring only a relatively small buffer in memory (8 − 16 Kbytes). If it is crucial that there are no gaps in the signal record, the 'double-buffer' technique can be used to maintain ADC sampling at the same time as data is being written to disc. This method is most conveniently implemented using DMA (Section 2.2.3) to transfer ADC samples from the laboratory interface into the temporary storage buffer. The DMA controller can be made to automatically re-cycle back to the start whenever the buffer is full, with this sequence continuing indefinitely until the controller is instructed to stop. As alternate halves of the memory buffer are filled they can be written to disc while ADC samples continue to be stored in the other half. No gaps will occur in the signal record as long as the contents of each half-buffer can be completely transferred to disc before it is required again.

Using this method continuous records containing 10 − 15 million samples can be collected, limited only by the storage capacity of the disc. Sampling rates in the range 27 − 70 kHz (depending on the interface), can be achieved with AT computers, 30 msec average access time hard discs, and laboratory interfaces which support DMA data transfer. To achieve these rates, care must be taken to optimize disc performance. To make the most efficient use of available disc space the MS-DOS operating system uses a non-contiguous file structure. If necessary, a file is split into parts allowing it to fill small vacant portions of the disc between existing files. File fragmentation must be avoided since moving from one part of the disc to another significantly prolongs the time taken to read or write a severely fragmented file. Disc optimizer programs exist which shuffle the files about on the disc, squeezing out all the small gaps, and combining the free space into a single large region. A file created after this process and before any other files are deleted will be unfragmented.

5.1.3 Synchronized signal collection

It will be apparent that if a very brief signal occurs only infrequently, continuous collection is very wasteful in disc storage space. For example a signal lasting less than 10 msec which might occur only once a second or less would utilize only 1% of a continuously collected data file. In such cases it is preferable to synchronize digitization to the occurrence of the signal and collect a short record only in the region of the signal. This is a straight forward task when the experimental signal is evoked by an external stimulus from which a synchronization pulse can be derived to trigger the start of digitization.

Spontaneously occurring signals, however, present a problem in that they lack any synchronization signal which can be used to initiate digitization. It is sometimes possible

Computer analysis of electrophysiological signals

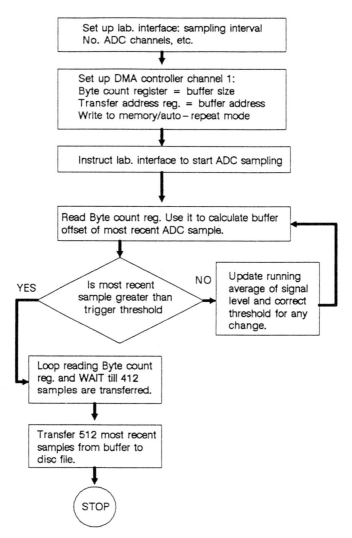

Figure 6. Flow chart of a spontaneous signal detection and collection algorithm. The PC DMA controller is used to fill repeatedly a 1024 sample memory buffer. To determine the presence of a signal, samples are compared to a threshold level which is adjusted for baseline drift using a running average.

to use the signal itself as the trigger but this inevitably means that all pre-trigger information is lost including part of the leading edge of the signal. A solution to this problem is to continuously digitize the analogue signal, storing in memory the 1024 most recent ADC samples. A software signal detection routine, running, in real time, in parallel with digitization, is then used to inspect the data being placed into the sample buffer. When a signal is detected a snapshot is taken of the contents of the buffer and stored in a disc file. This method allows pre-trigger samples to be included in the stored signal record. A signal is deemed to have been detected when an ADC sample exceeds a pre-determined trigger level relative to the signal baseline. To compensate for slow

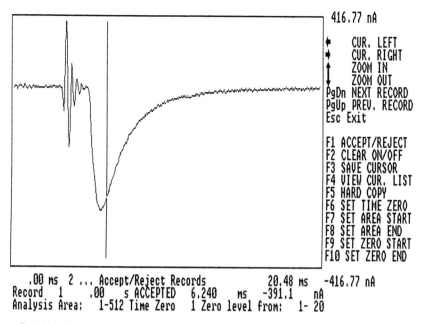

Figure 7. A 512 point signal record displayed on screen for inspection and validation. The user can change the display using function keys defined in the menu on the left of the display. Records can be marked as 'accepted' or 'rejected' using key F1.

drift in baseline a running average of the incoming signal can be calculated and used to adjust the threshold level to keep it at a fixed offset from baseline. The signal detector must be able to process the incoming stream of ADC samples at least as fast as they are acquired which requires that the routine be coded using asssembler language and carefully optimized for speed of execution. All calculations should be performed using integer arithmetic, where possible using only the fast internal CPU data registers. This task can be greatly simplified and higher performance achieved if ADC samples can be transferred from the laboratory interface into the data buffer using DMA. A simplified flow chart for the signal detection routine is shown in *Figure 6*.

5.2 Inspection and processing records

5.2.1 *Inspection and validation of digitized signals*

An often neglected aspect of the use of computers to analyse signals is that once the signal is stored within the computer it can be rather inaccessible to the user. Computers can sometimes be a mixed blessing, greatly enhancing the potential for quantifying data but drastically reducing the ability of the user to inspect it. Older methods such as recording on a pen chart or on 35 mm photographs of an oscilloscope screen while being slow to produce and analyse were easy to inspect. A computer analysis program may produce a result no matter how absurd the signal fed into it and it is not always possible to tell that an error has occurred from the final results. It is therefore necessary to visually inspect the whole of the signal record and to be able to exclude the invalid or corrupted parts from the analysis.

The most convenient way of inspecting a digitized signal record is to select a portion of it and show it graphically on the PC display screen. Since this operation might have to be performed by the user many thousands of times it is important to make it as fast and as easy as possible. *Figure 7* shows an example of one possible design of a display screen for signal inspection. The user controls the display of signals by means of the function keys F1 to F10 and the cursor keypad The screen can be divided into three areas.

(i)   A region showing the signal record being inspected.
(ii) A menu defining the operation of the function keys.
(iii) A data area showing cursor position, record number, etc.

In this program, collections of individual signal records containing 512 sample points are stored in a disc file. By pressing the F1 key, the user can set the status of the record to either 'accepted' or 'rejected' from further analysis. The module also allows the numerical value of any sample point within the record to be inspected by means of a movable cursor and a table of such values to be created and stored on an ASCII text file. This particular design makes extensive use of the standard function keys available on the IBM PC, F1−F10 and the cursor keypad. It was chosen because these keys can be found on all compatible PCs although their particular location may vary.

5.2.2 Signal averaging

One of the advantages of recording signals with a computer system is the capability of enhancing the signal. Many electrophysiological signals have a poor signal-to-noise-ratio which can be improved by taking the average of a series of records. To illustrate the use of averaging I have taken as an example some recordings of miniature end-plate potentials (MEPPs) from frog skeletal muscle. MEPPs are small signals, approximately 1 mv in height, with a poor signal-to-noise-ratio which can be improved by averaging.

To generate an average MEPP, a data file of MEPPs is collected using the spontaneous signal detection procedure, discussed in Section 5.1.3. Since it is important to ensure that each signal record in the series is truly typical of the phenomenon under study, the operator then makes qualitative judgements on the suitability of individual records for inclusion in the average. This may be as simple as the removal of obvious interference signals such the starting or stopping of a motor, or, occasionally, two MEPPs occur at the same time and partially superimpose, distorting the time course of the signal. Other cases are more subtle. Although more than 95% of the recorded MEPPs have a fast rise time of less than 1 msec, a small proportion of MEPPs have slow rise times and peak values greater than normal. The sizes and time courses of these giant MEPPs lie completely outside the distribution of the normal MEPPs and their is no reason to assume that they are even produced by the same underlying physiological process. It is therefore common practice to exclude such records from the average. This is not to say that they are not worthy of study only that this must be separate from the analysis of the more common MEPPs. *Figure 8a−f* shows a selection of the records in the series, typical both of those accepted and rejected.

Each MEPP is digitized into signal records of 512 ADC samples. To create the average

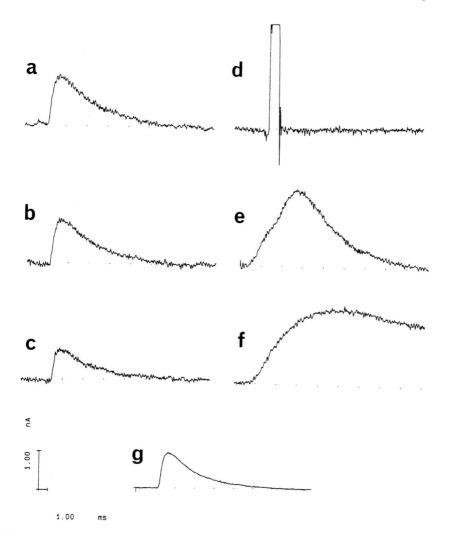

Figure 8. Signal averaging. A poor signal to noise ratio can be improved by averaging a series of signal records. (a)–(c) Records of typical MEPPs accepted for inclusion in average. (d)–(f) Records considered as unsuitable and rejected from average. (g) Average of 17 records similar to (a)–(c).

each sample point must be summed with the corresponding point in the other records. The average is then obtained by dividing the 512 sums by the number of records in the series, that is

$$\text{MEPP}_{\text{AVG}}(i) = \sum_{j=1}^{N} \frac{\text{MEPP}(i,j)}{N} \quad (i = 1,512) \tag{1}$$

where N is the number of records in the series, i is a sample point, j is a record number, and $\text{MEPP}(i,j)$ is sample i of record j. Equation (1) assumes that each MEPP is located in exactly the same place in each signal record. If this is not the case, it is necessary to identify the position of the signal in the record and then shift it to align with the same position of the average. With MEPPs it is usual to choose the mid-point of the leading edge of the signal as the alignment point. Note that it is important not to use the MEPP peak as an alignment point since this produces a small spike artefact at the peak of the average, caused by correlation between the signal and the background noise. *Figure 8g* shows the average created from 17 records by the above procedures.

5.2.3 *Leak and capacity current subtraction*

In the same way that a series of signals can be added together to enhance an electrophysiological signal, it is often useful to be able to subtract one signal record from another. A cell membrane almost always has a multiplicity of different species of ionic channels with different ionic selectivity, rates of opening and closure and voltage-sensitivity. Detailed study of any particular type of ion channel requires that the others be removed or blocked. This is possible to a large degree by adding pharmacological blockers such as Tetrodotoxin (sodium channel), Curare [nicotinic acetylcholine (ACh) channel], or by replacing certain ions in the physiological salt solution with impermeant ones (e.g. replacing Cl^-, with SO_4^{2-} to eliminate a Cl conductance). Even after such treatment however a certain amount of trans-membrane conductance often still remains. Current through this pathway is superimposed on the current under study complicating its interpretation.

However, in certain circumstances, it is possible to digitally remove this ionic leak current from signal records. To illustrate the process of digital leak current subtraction, a hypothetical record of a typical voltage-activated ionic current has been generated using a computer simulation (the Hodgkin–Huxley equation for a voltage-clamped current). *Figure 9a* shows the current recorded in response to a 25 mV voltage step. Three separate current components are added together within the signal.

(i)   A transient capacity current due to the abrupt change in cell membrane potential at the onset of the voltage pulse.
(ii) A voltage and time-dependent ionic current, the rising phase of which is barely discernible from the capacity current, which then decays exponentially to a steady-state towards the end of the voltage pulse.
(iii) A linear ionic leak current which is neither voltage- nor time-dependent.

Records of this sort are difficult to interpret. The relative proportions of components (ii) and (iii) making up the steady-state current at the end of the pulse cannot be determined. Neither can the rising phase of the current of (ii) be separated from (i).

Although, the leak current cannot be removed from this record, it is often possible to find conditions where another record can be collected containing only the leak and capacity current, such as shown in *Figure 9b*. It is therefore possible to isolate component (ii) of the current by digitally subtracted *Figure 9b* from *9a*, after scaling *9b* to account for any difference in the size of their respective voltage-steps. Leak-only records can be acquired in different ways depending on the type of experiment, by choosing a voltage-step to a potential which does not activate the voltage-sensitive current, by blocking the ion channel with a drug, or by removing a drug required to activate the current.

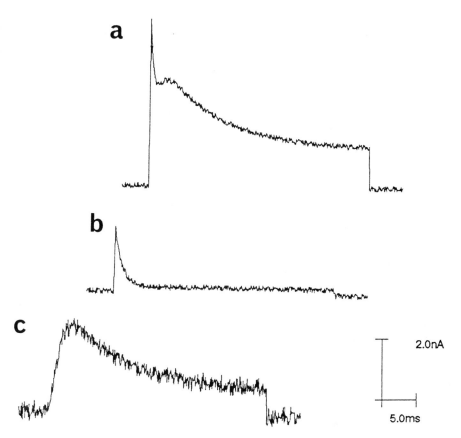

Figure 9. Linear leak current subtraction. (**a**) Simulated voltage-clamp current consisting of capacity transient current, voltage activated ionic current, and linear leak current (25 mV voltage-step). (**b**) Leak current only record (10 mV step). (**c**) Record (**a**) with leak and capacity current subtracted (4 leak records, scaled by 2.5).

If the voltage-step of the leak current record is substantially smaller than the test record it also necessary to average a series of leak records before subtraction, since any scaling up of leak current also magnifies its background noise. *Figure 9c* is the result of the average of four leak records, scaled by 2.5 (test/leak voltage ratio, 25 mV/10 mV), and subtracted from the test records. To prevent an excessive increase in background noise, the number of leak records averaged should be equal to, at least, the square of the scaling ratio.

5.3 Automatic measurement of signal records

The aim of digitizing an electrophysiological signal is usually to quantify some aspect of the amplitude or time course of the signal. Amplitude measurements are often made relative to a region of the signal record defined as the baseline level rather than to an absolute zero level. This procedure corrects for any alteration of the baseline from one record to the next. Two regions in the record must therefore be defined to complete a measurement, the baseline region and the measurement region itself. It is useful to take the average of a series of samples within these regions to minimize the effects

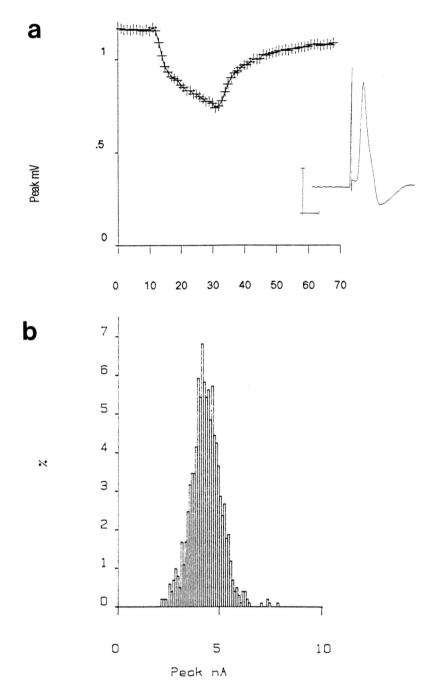

Figure 10. Quantitative record measurements. (**a**) Change in a measured parameter with time. Peak amplitude of a series of 68 records (collected at 20 sec intervals) of compound action potentials from frog sciatic nerve. Amplitude is depressed by addition of local anaesthetic at the 10th record. (**b**) Statistical distribution of a parameter. Histogram of peak amplitudes of 1000 miniature end-plate currents.

of background noise on the measurement. This is especially important with peak measurements. A number of different parameters can be obtained from the measurement region.

(i)   The average signal amplitude within the region.
(ii) Positive or negative peak amplitude.
(iii) The area underneath the signal.
(iv) Signal rise time from 10 to 90% of peak.

The position of the beginning and end of the baseline and measurement regions can be set using a movable cursor superimposed on the screen display to indicate the desired position then pressing a function key. In *Figure 7*, for instance, keys F7 and F8 set the measurement region and F9 and F10 set the baseline region. The ability to define a specific measurement region can be used to clearly separate the part of the record containing the signal of interest from that containing a stimulus artefact.

Once the region of interest with the signal record has been defined the actual calculation of parameters (i) to (iv) can be performed automatically for many thousands of records. The subsequent analysis of these results falls into two general categories, the analysis of the changes in a measured parameter with time or the statistical distribution of the parameter. *Figure 10a* illustrates an analysis of parameter changes with time. The peak amplitude of a series of 68 digitized compound action potentials, recorded from frog sciatic nerve, is plotted versus record number (collected every 20 sec) within the data file. The aim of this experiment was to show the effect of a dose of local anaesthetic, added at the 10th record and washed out at the 30th. A typical signal record from this experiment is shown inset with the measurement region indicated. On the other hand when the value of a parameter is fluctuating randomly but shows no trend with time, the statistical distribution is more informative than the time course. *Figure 10b* is a histogram of the peak amplitude of a series of 1000 MEPCs recorded from a voltage-clamped snake neuromuscular junction.

5.4 Curve fitting

A great deal of information relating to the mechanisms underlying the function of ion channels can be derived by studying the effects of membrane potential, temperature and pharmacological agents on the time course of electrophysiological signals. This involves choosing a hypothetical model for the process under study, then deriving a mathematical equation, commonly consisting of the sums or factors of exponential functions, which describes the time course of all, or a portion of, the signal. For example, in *Figure 11* a voltage step from 0 to 100 mV has resulted in a current signal which, after an initial abrupt rise, relaxes exponentially to a steady-state level (voltage-clamped Chromaffin cell, ACh activated current). The exponential region of the signal can be described by the following equation:

$$I_f(t) = A \times \exp(-t/\tau) + ss \qquad (2)$$

where $I_f(t)$ is the current time course, ss is the steady-state current level, and A is the amplitude of the exponential relaxation, measured from the initial current (where $t = 0$) to the steady-state, τ is the time constant of relaxation. Equation (2) is the general model

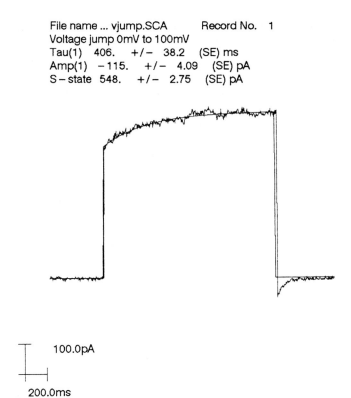

Figure 11. Curve fitting. An exponential function fitted to the time course of the change in voltage-clamped ionic current produced by a step change in holding potential from 0 to 100 mV (Bovine Chromaffin cell, ACh activated current). Equation (2) has been fitted to a section of the record with a non-linear iterative curve fitting algorithm using the Levenberg−Marquadt method. The best fit parameters, A, τ and ss, and their standard errors are printed above.

which we assume is a good representation of the actual current. A curve fitting method must now be applied to determine whether the model does in fact fit the data, and if it does, find the quantitative values of A, τ and ss which provide the best fit.

As for the measurement of signal amplitude, the region to be analysed within the signal record along with a signal baseline region is defined by the user. A fifth point must also be added to define where the exponential curve is deemed to begin [$t=0$ in equation (2)]. With this information the exponential segment of the signal can be extracted and passed to the curve fitting procedure.

It is common to use the quantity known as χ^2 (chi-squared) as a measure of 'goodness of fit', calculated by the equation:

$$\chi^2(\tau,A,ss) = \sum_{i=i_1}^{i_2} \frac{[(I_d(i)-I_z)-I_f(i-i_0)]^2}{\sigma^2} \qquad (3)$$

i_1 and i_2 define the beginning and end of the region of the signal record containing the data to be fitted. i_0 defines the point at which the exponential function begins, I_d is the data point from the signal record, I_z is the signal zero level, I_f is the fitted equation, and σ is the standard deviation of the data points relative to the fitted equation $I_f(i - i_0)$. Analytical solutions which minimize equation (3) can be derived for certain types of function such as the straight line, exponential function decaying to zero, and those described by polynomial equations. But for most other functions, the minimum can only be obtained by an iterative numerical curve fitting process, in which χ^2 is repeatedly calculated for a series of trial values of the function parameters A, τ and ss in the case of equation (2) until the minimum value of χ^2 is found. The design of a good curve fitting algorithm is a difficult process requiring a detailed knowledge of numerical analysis. Fortunately, general purpose algorithms are available as part of many mathematical software libraries (NAG PC50 Software Library and ACM Algorithms Distribution Service). The programmer need only supply a subroutine to calculate the function to be fitted. The algorithm used here is a version of the Levenberg – Marquadt method, written by K.M.Brown of the University of Cincinatti. Further details of non-linear curve fitting algorithms, including source code, can be found in chapters 10 and 14 of reference 5, and algorithm AS47 of reference 9.

No matter what fitting method has been used it is important to always superimpose the curve of the best fitting function over the actual signal record, as shown in *Figure 11*, in order to determine whether the the results appear plausible. *Figure 11* also shows the parameter values for the best fit, along with their standard errors calculated from a covariance matrix. These standard errors can be considered as indicators of the confidence limits of the given parameter values. They do not, however, give any indication of the goodness of fit. The user must decide whether any residual deviation of the fitted curve from the data is due to background noise only, or an indication that the chosen function is incorrect. If, for instance, the fitted curve exceeded the data points at the beginning and underestimated at the end of the fitted segment, even slightly, but consistently for all of the records analysed, then this would suggest that perhaps the sum of two exponentials might be a better choice of fitting function. Results obtained by curve fitting should therefore be treated with caution and one should take pains to convince oneself that the best fitting equation has been chosen.

5.5 Analysis of patch clamp recordings

5.5.1 *Single channel analysis*

The discovery by Neher and co-workers that a $1-2$ μm diameter fire-polished tip of a glass micro-pipette could, under appropriate conditions, form a very high resistance 'gigaseal' with a patch of cell membrane has greatly increased the scope of electro-physiological recording from a wide range of cells. In particular, it has allowed the study of the fluctuations in current passing through single ion channels isolated within the patch. Ion channels fluctuate between a closed, non-conducting, state and one of a small number of discrete open states of constant conductance. The time spent in each state is a value randomly selected from an exponential distribution of times. Due to the stochastic behaviour of the signal very little can be inferred from the duration of a single opening or closure. Analysis requires the measurement of hundreds and often thousands of open or closed times in order to calculate the mean or the distribution

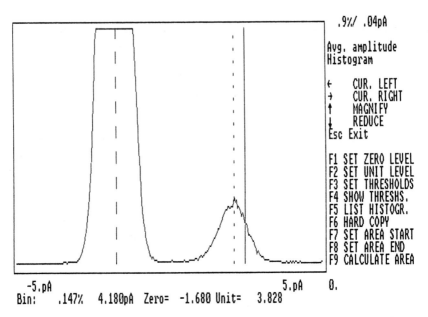

Figure 12. Current amplitude distribution as displayed on screen from the single-channel analysis program. Zero and unit current levels, for transition detection can be set using cursor and function keys.

of times. Single-channel recordings fall into two broad classes which are analysed in different ways.

(i) *Ion channels in a steady state fluctuating between open and closed states for long periods.* These channels are usually activated by an agonist such as GABA, acetylcholine, or glutamate. The preferred method of analysis is to digitize a continuous record of the signal lasting for several minutes, store it on disc, and then to inspect and analyse the record.

(ii) *Ion channels which open in response to a stimulus then rapidly inactive.* These channels are usually voltage-activated, remaining in the closed state until an abrupt change in membrane potential produces an opening, or a short burst of openings, before the channel enters an inactivated state. Such channels must be returned to the original holding potential before they re-activate and become capable of opening again. The Na channel which underlies the nerve action potential is a typical example. Rather than a single long continuous record this type of channel should be digitized as a series of short records synchronized to the voltage-step and of the same duration.

5.5.2 *Current amplitude distribution*

It is useful to start the analysis of the single-channel record by generating a histogram of the current amplitude which can then be used to obtain estimates of the zero current and channel unitary current levels. To calculate the amplitude histogram the total ADC range of the digitized signal record (i.e. 4096 for a 12-bit ADC) is split into equal sized histogram bins. Blocks of 512 ADC samples are copied from the digitized signal record file into an array in memory. The value of each point, divided by the number of bins (i.e. 512), is used as an index into the histogram array and the appropriate entry incremented. *Figure 12* shows a screen display of an amplitude histogram compiled in this way. Such histograms typically have two main peaks, often with approximately Gaussian shape, one at the zero current level and the other at the unit current level. A cursor can be moved along the display and the position of these levels entered into the program by pressing a function key. In *Figure 12*, the positions chosen for the zero and unit current levels are indicated by dashed and dotted vertical cursors.

The total time spent in both the open (T_{op}) and closed (T_{cl}) channel states can be directly calculated from the amplitude histogram by adding up the number of ADC samples found within the range of each of the peaks in the histogram then multiplying the total by the sampling interval. If the peaks are well separated, the beginning and end of each peak can be easily defined on the screen with the cursor. The mean open channel probability can then be calculated from the equation:

$$P_{op} = \frac{T_{op}}{(T_{op} + T_{cl})} \qquad (4)$$

The amplitude histogram while being a simple and unequivocal measurement has a number of limitations. If the zero current level of the digitized signal is drifting the peaks of the amplitude histogram will be distorted. In such circumstances the signal amplitude histogram must be measured relative to the local signal zero level which requires some means of determining when the channel is closed. Similarly, while the total time spent in each state can be calculated from the amplitude histogram, the mean time cannot, since no information is provided on the number of channel openings and closures. Further analysis therefore requires that the transitions between the open and closed states be detected and the individual times spent in each state measured.

5.6 Channel open/close transition detection

The heart of any single channel analysis program is its method of determining, from the digitized current record, when the channel changes state, in order to calculate a list of the times spent in each state. This in fact turns out to be more difficult than it might at first appear and there is still a certain amount of controversy over what is the best means of doing so. The first problem is that channel openings and closures may be of a duration of the same order or shorter than the response time of the patch clamp recording system. Poor signal-to-noise-ratios often require that the recorded signal be low-pass filtered at between 1-4 kHz while some channel openings or closures can be as brief as 10-100 μsec. Low-pass filtering distorts the time course and reduces the size of such brief events (see Section 3.4), eventually so that they are indistinguishable from the system noise and cannot be detected. This has two serious consequences. Firstly, the distribution of durations of the brief events will be distorted. Secondly, missed events will result in the distortion of the complementary distribution of states even though they may be of long duration, that is missing a brief opening will produce a double-sized closed state and *vice versa*. Another problem is that some species of channels have more than one open state, each with a different conductance.

5.6.1 50% Threshold transition detection

The simplest, and probably, the most commonly used approach to detecting channel transitions is to set a threshold at a level half way between the zero current level and the unit current level, as determined from the amplitude histogram. When the signal is below this level the channel is deemed to be closed and open when above. Starting at the beginning of the digitized data file, each ADC sample is compared against the threshold to determine whether the channel is in the open or closed state. A list of the times spent in a state is determined by counting the number of consecutive samples before the threshold into the other state is crossed, then that time is written to a list file. In the program described here the list file also contains information on where the digitized record of the channel opening or closure is located within the digitized data file, for later inspection by the operator. The method can be refined slightly to keep track of an average of the zero current level and to adjust the threshold to compensate for any slow drifting of the signal baseline. While being simple to implement the 50% threshold method has two serious deficiencies. The method is not particularly sensitive to brief events since it will fail to detect any event which has been reduced in amplitude to less than 50% of the unit current, even though such events are still clearly evident when displayed on screen. Also sub-conductance levels close to 50% of the designated unit conductance in size, can produce long series of false transitions caused by the background noise crossing back and forth across the threshold. Nevertheless, the 50% threshold method is a fast automatic means of detecting channel transitions and, as long as these limitations are respected, it can be applied to a wide variety of existing species of channels.

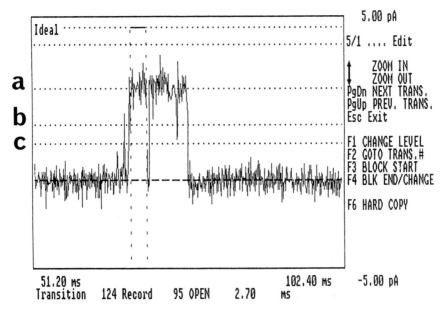

Figure 13. Inspection and validation of results of transition detection procedure. A postulated 'open' state is displayed on screen with the actual digitized current record for comparison. The user can change the state with F1. Zero current level indicated by dashed line. Unit current level indicated by (**a**), transition thresholds by (**b**) and (**c**). (Note that a two threshold detection method has been used.)

In the single-channel analysis program described here, transition detection is implemented using a similar method, except that two thresholds are used, placed as close to the zero and unitary current levels as the background noise will allow (see *Figure 13*). When the signal is below the lower threshold the channel is deemed as 'closed', when it is above the upper threshold it is 'open', and when it is between the thresholds it is in an ill-defined sub-conductance state. This methods provides improved detection of brief events compared to the single threshold method and also prevents some of the errors introduced by the presence of sub-conductance levels. It cannot, however, be regarded as a complete solution to the detection and measurement of multiple sub-conductance states.

5.6.2 *Inspection of transitions*

When using automatic transition detection it is crucial that the operator can inspect and, if necessary, change some of the program's results. The digitized record may, for instance, contain short bursts of noise related to transient breakdown of the gigaseal, or periods where more than one channel is open simultaneously, which have to be eliminated from the analysis. Each discrete open or closed duration in the list can be displayed on the screen as shown in *Figure 13*. A portion of the digitized signal record containing the event is displayed with it's duration and designated state indicated by the idealized trace above. The operator can then alter the state by pressing key F1 to cycle through the possible states: 'closed', 'open', and 'rejected'. The list of state durations is then checked and sequences of list entries with the same state are combined into single entries.

5.6.3 *Time course fitting*

Alternative methods to the 50% threshold exist which while being substantially slower in execution and more difficult to implement, are more robust and probably extract as much information from a single channel current signal as is possible. A method developed by David Colquhoun known as 'time course fitting' substantially improves the estimation of brief events. The details of this method are described along with other useful information in Chapter 11 of reference 2, but a summary will be described here. In Colquhoun's system (based on a PDP11 minicomputer) the digitized signal record is scanned by the program until the signal exceeds a threshold set close to the zero current level, indicating that a channel opening or the beginning of a burst of openings has been found. That segment of the digitized signal record is then displayed on an oscilloscope screen. To determine the duration of the opening, the operator adjusts the alignment of two pulse steps representing the rising and falling edges of a low-pass filtered, ideal channel opening until their sum exactly superimposes on the actual opening. The time course of the step pulses depends on the degree of low-pass filtering that has been applied to the signal and may vary between experiments. They can be obtained either as averages of the opening and closing transitions of a series of long well-defined openings or by injecting square test current pulses into the input stage of the patch voltage-clamp.

The prime advantage of time course fitting is that it can compensate for the effects of the limited bandwidth of the patch clamp, allowing more accurate estimation of the

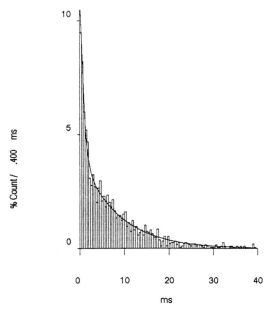

Figure 14. Histogram of 1888 channel open state durations, derived by transition detection. Best fitting two exponential probability density function is superimposed with time constants and areas printed above.

duration of very brief events. The method has been used to measure events as short as 40 μsec (6). Another advantage is that, unlike the threshold methods, it can be easily extended to analyse channels with multiple sub-conductance levels (7). It's main disadvantage is that the process of visually superimposing the ideal time course onto the digitized signal is a slow one, even when assisted by an iterative curve fitting routine as in Colquhoun's latest versions of the software. It is also dependent on the judgement of the operator as to what is, or is not, a true sub-conductance level, although this may be regarded as an advantage.

5.7 Open and closed time histograms

The distribution of the times spent in a particular channel state provides information on the kinetics of the underlying molecular processes and usually takes the form of the sum of one or more exponential functions. *Figure 14* shows a histogram of the open times from a series of 1888 channel openings, obtained using two threshold transition detection. A probability density function (pdf), the sum of two exponential functions, has been fitted to the histogram, using the curve fitting method described in Section

5.4. This method for fitting the pdf is suitable in cases, such as this, where the exponential time constants are sufficiently similar in value, so that the whole of the pdf can be adequately included within the range of a single histogram. On the other hand, distributions are encountered which require three separate exponential components with time constants ranging from tens of microseconds, through milliseconds, to hundreds of milliseconds. A histogram range cannot be found for these distributions which includes the longest intervals while maintaining a bin size small enough to resolve the shortest intervals. In such cases, the pdf should be fitted using the 'maximum likelihood' method which fits the pdf using directly from the list of state times, without calculating a histogram. See reference 2 for a discussion of this method and further details of the interpretation of open and closed state distributions.

5.8 Noise analysis

The ionic current recorded from a whole cell is no more than the sum of the currents passing through each of a population of individual ion channels. Even under steady-state conditions, the random openings and closures of the channels cause the whole cell current to fluctuate about the mean level. Although the rectangular current steps of the single channels are no longer discernible, the frequency distribution of this current 'noise' contains useful information about the channel kinetics which can be extracted by the method of noise analysis. Also, the amplitude of the current fluctuations relative to the mean level can be used to estimate the conductance of a single ion channel. Although noise analysis has in many respects been superseded by single channel recording it is still of some value as a quick means of estimating channel kinetics from whole cell patch clamp experiments and from tissues which are not suitable for patch clamping.

The central measurement in this method is the calculation of the power density spectrum of the fluctuating current signal. The power spectrum is a measure of the frequency distribution of the variance of the signal about its mean value and is calculated by applying the fast Fourier transform (FFT) algorithm to the digitized record of the signal. To summarize, it can be shown that a signal waveform of any desired shape can be synthesized by the summation of a series of sine waves with appropriate amplitudes, frequencies and relative alignment (phase). It follows that any signal can be represented equally well as either a spectrum of frequency points (frequency domain) or as a series of points in time (time domain). Although it is possible to perform all of the noise analysis calculations in the time domain certain features of the frequency domain representation of the signal make it more suitable. Firstly, it is difficult to separate the ionic current noise from the background noise of the measurement system using time domain methods only, but as will be seen, it is a simple matter in the frequency domain. Secondly, the existence of the FFT algorithm makes the calculation of the power spectrum much more efficient than of the equivalent time domain measurement, the autocorrelation function.

5.8.1 *The fast Fourier transform*

The fast Fourier transform is an algorithm for the conversion of a signal from the time domain to the frequency domain, and is termed 'fast' because it exploits certain redundancies in the calculation to greatly reduce the number of computations required. One restriction in the use of the FFT is that the number of sample points in each record must be a power of two, such as the 512 point FFT described here. The source code for the FFT algorithm is available in many computer languages, a well known one being that devised by D.M.Munro in FORTRAN (algorithm AS97 in ref. 9) Further details of the theory and use of the FFT algorithm can be obtained in references 5 and 8. The number of sample points (N) and the digital sampling interval (dt) determine the frequency range of the power spectrum, by the formulae:

$$F_{low} = 1/(Ndt) \tag{5}$$

$$F_{high} = 1/2\, dt \tag{6}$$

Any signal frequencies that lie outside the range F_{low} to F_{high} will appear as aliased frequencies (see Section 3.2.1) within the power spectrum and therefore the cut-off frequencies of the analogue high-pass and low-pass filters are usually set close to these limits. The FFT algorithm does not directly calculate the power spectrum but rather produces a spectrum of $N/2$ points, equally spaced in frequency, expressed in complex arithmetic which contains both amplitude and phase information. The power spectrum, itself, which only contains amplitude information, is calculated by the further step:

$$P(f) = \frac{2\ \{\mathrm{Re}[\mathrm{FFT}(f)^2] + \mathrm{Im}[\mathrm{FFT}(f)^2]\}}{N\, dt} \quad (f = 1..N/2) \tag{7}$$

where Re and Im indicate the real and imaginary parts of the FFT.

5.8.2 Calculation of the power spectrum

Figure 15 shows the significant stages in the process of calculating the power spectrum of ionic current noise. Recordings of ACh induced end-plate current are used as an example of typical data. Initially, a series of 512 point signal records of the current fluctuations (*Figure 15b*) in the presence of ACh are digitized and stored on file. Unlike the previous records shown, these have been high-pass as well as low-pass filtered to remove all signal frequency components (including any steady-state DC level) that lie outside the range of the spectrum to be calculated. For the reasons mentioned in Section 3.2.2, a Chebyshev filter with a high roll-off rather than a Bessel filter has been used. A similar series of records (*Figure 15a*) is also collected of the background noise recorded in the absence of the agonist. The records are visually inspected and those containing artifacts such as MEPCs are excluded from analysis. The power spectrum is then calculated for each 512 point record (*Figures 15d*, ACh and *15c*, background), by means of the FFT. Before transformation, the edges of the data are tapered off using a 10% cosine bell data window in order to minimize spurious side-lobes produced by any sharp peaks in the spectrum (see ref. 8). This process requires that the resulting spectrum be multiplied by a correction factor (1.143). The power spectrum calculated from a single 512 point record is, in itself, of little value since each frequency point has a very large standard error. It is, however, possible to make the spectrum interpretable by taking the average of the results of a large number (e.g. 32) of individual FFTs. *Figure 15e* shows such averaged spectra containing the background noise spectrum and the pure ACh-induced current spectrum obtained by subtracting the background spectrum

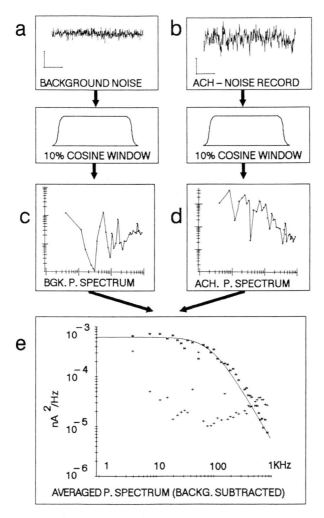

Figure 15. Noise analysis. The calculation of a power spectrum, and fitting of a Lorentzian curve (equation 8), from steady-state end-plate current noise invoked by ionophoretically applied ACh. (**a**) 512 sample record of background current, (**b**) record after application of ACh (calibration 1 nA, 50 msec). (**c**) power spectrum calculated by 512 point FFT from one background record, (**d**) spectrum from ACh noise record. (**e**) Average of 28 ACh power spectra after subtraction of average background spectrum and scaling by cosine window correction factor. Best fit Lorentzian curve superimposed, $S_0 = 6.2 \times 10^{-4} nA^2$, $\tau_{op} = 1.7$ msec, $\gamma = 33.6$ pS.

from the raw ACh current + background spectrum. Note that the spectrum is plotted in the standard fashion of log power (nA^2) versus log frequency (Hz).

5.8.3 *Lorentzian curves*

The shape of the power spectrum in *Figure 15e* is typical of that produced by the activity of a population of independently gated ion channels. The power is constant up until a cut-off frequency above which it falls off proportional to the square of the frequency.

It can be shown (see ref. 11) that current noise produced by a population of ion channels with exponential distributions of open and closed states might be expected to produce such a spectrum and that it should conform to the equation:

$$S(f) = \frac{S_0}{[1 + (f/f_h)^2]} \quad (8)$$

where S_0 is the power level at low frequency limit of the spectrum and f_h is the half-power frequency. Equation (8) has been fitted to the spectrum in *Figure 15e*. If certain assumptions hold, in particular that the channel only has a single open conductance, zero conductance in the closed state and if only a small fraction of the available channels are in the open state at any given time then the single channel conductance (γ) and channel open time (τ_{op}) can be calculated using the following equations.

$$\gamma = \frac{S_0}{2\pi \times f_h(I_m)(V - V_r)} \quad (9)$$

$$\tau_{op} = \frac{1}{(2\pi \times f_h)} \quad (10)$$

where I_m is the mean steady-state current, V is the membrane potential, and V_r is the ionic current reversal potential.

In general, noise analysis is most easily applied in experimental situations where a steady-state ionic current can be induced by the application of an agonist and does not significantly desensitize or inactivate within the period of recording. The method can be extended to analyse channels with more complex kinetics where the spectra consist of more than one Lorentzian component to voltage as well as agonist activated channels, and to ionic currents which inactivate. Details of the method and its underlying theory can be found in references 10 and 11.

5.9 Command voltage pulse generation

So far, only the computer analysis of electrophysiolgical signals after they have been acquired has been discussed. However, the computer can also be of assistance while performing the experiment. In this role, probably the most useful function that a computer can play is the generation of series of voltage pulses for the control of a voltage-clamp. Although series of single step voltage pulses can be produced by conventional analogue stimulators, complex patterns are difficult to set up and adjust especially when produced at rates greater than one pulse per second. It is however quite a straightforward task to connect a D/A converter output from the laboratory interface to the command voltage input of the voltage-clamp and to produce a programmed pulse pattern using software.

Figure 16 shows the operation of a typical voltage generator program. It has been set up to produce a series of 10 two-step voltage pulses at one second intervals with the amplitude of the second pulse incremented by 100 mV between pulses. The upper half of the screen shows the pulse series that has been produced and sent to the D/A converter. The lower half shows the pulse programming procedure where pulse amplitude, duration and inter-pulse increments can be entered. A variety of such programs can be defined in this way, stored on file, and invoked later at the touch of a button. Each voltage pulse is displayed on the screen as it is sent to D/A output 0.

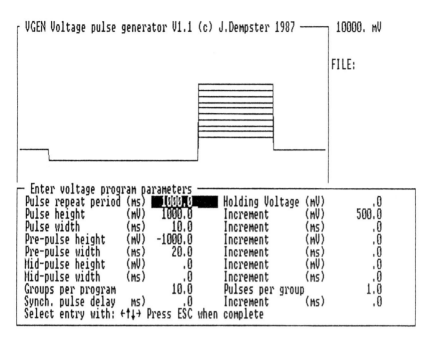

Figure 16. Screen display of a voltage pulse generator program. The program is shown just after the completion of a sequence of two-step pulses. The data entry table defining pulse amplitudes and durations is shown superimposed.

A 1 V amplitude, 1 msec duration synchronization pulse is also produced on D/A channel 1, which can be used to trigger a monitor oscilloscope or recorded along with the other signal on tape to indicate the start of the signal record.

5.9.1 Combined 'on-line' pulse generation and recording

Voltage pulse generation and signal digitization can be combined together into a single 'on-line' program for controlling and collecting data from an experiment at the same time. A particular advantage of this approach is that information derived from data analysis carried out during the course of the experiment can be used to select appropriate futher experimental procedures. It might, for example, be useful to compute and display the I/V curve of single channel currents before, during and after the alteration of the concentration of various ions in the bathing solution. The Cambridge Electronic Design 1401 interface is particularly suited for this purpose since its 6502 microprocessor can be programmed both to collect ADC samples and to output DAC values at the same time, to and from arrays within its on-board memory. To achieve the same effect with other interfaces such as the Data Translation DT28XX series two separate boards are required for each task.

5.10 'On line' versus tape

While 'on-line' data recording can eliminate the need for an expensive analogue FM tape recorder there are a number of reasons why it is better to continue recording on tape even if an 'on-line' system is available. The storage capacity of even the largest

Winchester disc is much less than that of analogue magnetic tape, In practice, 150 Mbytes of digital storage capacity are required to store the equivalent of a typical 1200 foot tape with two FM channels recorded at 7.5 IPS (5 kHz bandwidth). Recording all experimental data in digital form is likely to be much more expensive than in analogue form. Also the setting of the tape recording speed (and hence bandwidth) is a much less critical task than choosing the digital sampling interval since the high capacity of tape media allows a signal to be recorded at a relatively wide bandwidth then later filtered and digitized with a variety of sampling rates. Action potentials in cardiac muscle, for instance, often have a very fast initial component which requires sampling rates greater than 20 kHz for its study while the remainder of the signal lasting as long as 500 msec is better digitized at a 1 kHz rate. 'On-line' recording of such a signal would require approximately 40 Kbytes of storage per action potential (2 bytes at 20 kHz for 1 sec) and then a complicated process of digital filtering and data reduction to view the whole signal at an effective 1 kHz rate. Finally, while it is a simple matter to replay a tape and have the signal digitized and analysed by any suitable computer, it is not generally possible to do this with digitized data files which are specific both to type of computer disc formats and software used when it was recorded. Given the rapid advances made in both computer hardware and software, it is distinctly unwise to record large amounts of data in a form which might become obsolete within a few years.

5.11 Pulse code modulation recording

The main difficulty with FM recording has been the high cost of recorders and the relatively high tape noise level. In the last few years, however, a lower cost alternative has been found by adapting standard domestic videorecorders for recording electrophysiological signals. A standard VHS or BETAMAX video recorder is combined with a modified SONY PCM701 digital audio processor to provide two data channels with a DC−20 kHz bandwidth, a voice channel and a synchronization channel. The pulse code modulation recording method is used, essentially a hybrid technique where an analogue signal is digitized, in this case at a 44.2 kHz rate, encoded along with error correction information and recorded in the form of a video signal. On playback, the signal is decoded from the video signal, errors are corrected, and the analogue signal is reconstituted via D/A converters. The system can therefore be used in exactly the same fashion as an FM tape recorder. The main advantages of the technique are the high quality of the recording (16-bit accuracy) and low cost, especially of the recording media which are domestic video tapes. The required modifications to the Sony PCM701 are detailed in a paper by one of the orginators of the technique, Trevor Lamb (12), and are basically, the removal of filter capacitors to extend the device's frequency response to DC and the addition of a calibrated input voltage range.

6. DEVELOPMENT OF SCIENTIFIC SOFTWARE
6.1 How to obtain software

There are four main paths to obtaining data analysis software. A commercially available scientific package may be purchased, a software company may be commissioned to write customized software according to one's specifications, software may be obtained from another laboratory in the same field of work, or the software may be developed 'in-house'. Each approach has its own merits and problems, but the basic rule is that it is either going to cost money or it is going to cost time. The software described in this chapter has been developed 'in-house' at the University of Strathclyde and represents so far five man-years of effort. This is a fairly substantial commitment of time and resources and was initially justified because it supported the activities of a group of 4–5 experimental workers. More recently, the work has been generalized and extended to include laboratories working in the same field in other Scottish universities and the UK, and we have been fortunate to obtain grant support for this endeavour. It is currently our policy to make this software (both source and executable code files) freely available to other academic, non-profit making, laboratories while retaining the rights to market it to commercial organizations.

6.2 Software design goals

The design goals used in developing the software described here may be of interest to readers contemplating producing their own packages. From the outset the software was intended to be 'user-friendly', in that it was to be usable by experimental scientists who might have only a basic knowledge of the workings of the computer and none of programming. To this end the programs were designed with a menu-based rather than a command-driven user interface. Programs operated by command words entered at the keyboard require the user to remember a complex sequence of command names and arguments. Experienced computer programmers do not find difficulty in remembering long lists of esoteric words but novice, or occasional computer users, often perceive this as a daunting prospect, the MS-DOS command set being a typical example. On the other hand, one of the reasons for the popularity of the LOTUS 1-2-3 spreadsheet is its well-designed menu system. The PC keyboard with its function and cursor control keys also lends itself well to menu-based software.

Another choice was to create several customized programs for each distinct type of experiment rather than a single all-encompassing program. Different types of experiments require not only different forms of analysis but also the data files to be structured in different ways, making it difficult to design one program which could satisfy all possible applications. The program for the analysis of EPCs, for instance, required a data file split into a large number of individual signal records (one per EPC) while the single channel current program was best served by a single very large record. Also a single program capable of everything from signal averaging, power spectrum analysis, to single channel transition detection would have been hopelessly complex to operate by each user who might only have required a small subset of its features. Good program design therefore involves determining what are the desirable features to incorporate into each individual program and which to relegate to other ancillary programs.

6.3 Programming languages

The software described here has been written in the FORTRAN 77 language (Digital Research compiler) and in 8086 assembler (IBM or Microsoft assembler). The main reason for choosing FORTRAN 77 was the author's familiarity with the language but it has a number of merits in its own right. After BASIC, FORTRAN is probably the language that is best known by scientists and engineers. Therefore, a wide range of useful subroutines for numerical analyses exist in this language and can be easily obtained either by consulting a textbook containing the source code or by purchasing a scientific subroutine package. FORTRAN is a highly standardized language allowing programs to be easily ported from one machine to another as long as it has a FORTRAN compiler. The main drawback with FORTRAN is that it is not particularly elegant, lacking most of the block-structure commands (CASE.. WHILE.. UNTIL.. REPEAT..) which contribute to the elegance and readability of the Pascal language. However, due to FORTRAN's lack of structure (and almost to Pascal's excess) it is much easier in FORTRAN to produce a large and readily re-usable library of subroutines for manipulating arrays of numbers. FORTRAN and Pascal are 'compiled' languages and therefore to use them a source code file must be created with a text editor, that file is then converted to a binary (object code) file (by the compiler) and then the object code file is often joined together with other object files using a 'linkage editor' to finally produce the executable code file. On the other hand BASIC is commonly implemented as an 'interpreter' in which the lines of source code are translated into machine instructions (interpreted) one by one as the program is executed. The role of text editor and compiler is compressed together into a unified programming environment in which the user is unaware of these details. It is therefore much easier for the beginner to write and test programs in BASIC rather than in one of the compiled languages, but interpreted BASIC is much slower than any compiled language. A program originally written and tested in Microsoft BASIC however, can be later compiled separately using a BASIC compiler to produce a program which runs as fast as FORTRAN or Pascal. Conversely, the Turbo Pascal language, through using a very fast compiler with a built-in text editor comes close to providing the ease of use of an interpreter.

In the end, one has to admit that the choice of programming language is rather a matter of taste, as long as the would-be programmer knows the limitations of their choice. BASIC is ideally suited for small scale projects consisting of no more than 1000 lines of code, which have to be produced quickly by relatively inexperienced programmers. On the other hand, the total absence of any kind of language structure in BASIC, especially the lack of independent subroutines, makes the writing of large programs unmanageable. A better structured language such as FORTRAN, Pascal, or 'C' must therefore be used for large projects. Becoming proficient in such languages and in the wider aspects of program design however demands a much greater investment in time and effort.

6.4 Graphics virtual device interfaces

In many respects, the choice of programming language is much less important than of the software used for graphical display. The wide variety of graphics display codes and interfaces for display screens, printers and plotters discussed in Section 4, creates a problem when writing software to plot graphs on all these devices. The CGA, EGA and Hercules display adapters, for instance, all have different hardware device registers and arrangements of the screen memory. For this reason, programs making use of the screen graphics commands in Microsoft's GWBASIC or IBM's BASICA will only work with the CGA. The solution to this problem chosen by many suppliers of commercial software packages such as LOTUS 1-2-3 is to separate out the device specific parts of their software into device driver routines and to provide a selection of these for all commonly available display adapters. It can also pay to take this approach when developing one's own software. Software packages known as graphics virtual device interfaces (VDIs) are available which act as a buffer between a program and the display. Graphics are drawn on a virtual display screen using standard graphics commands with the details of translation into the device specific code being left to the device driver routine. The programs described in this chapter have been developed using Digital Research's GEM VDI. This package provides a range of software device driver for all of the commonly available graphical display and plotting devices, including, CGA, EGA and Hercules display screens, Epson and IBM dot matrix printers, Hewlett Packard plotters, and Apple and Hewlett Packard laser printers. When installed, GEM VDI behaves as an extension to the MS-DOS operating system with instructions being sent to it via a DOS software interrupt. A simple assembler language subroutine was used to pass graphical data point arrays from the FORTRAN program via this route to the VDI. Using this method, the same program code can be used to draw graphs on any attached device simply by selecting the associated device number. Details of GEM VDI and its interface to certain programming languages can be found in the Gem Programmer's Toolkit. Other similar packages include METAWINDOWS and WINDOWS developer's toolkit.

6.5 Some commercially available software

Commercially available software can be split into two categories, general purpose software and special purpose scientific software. The general purpose category includes products such as spreadsheets (LOTUS 1-2-3) and word processors (Microsoft WORD) which although primarily intended for the business market are equally applicable to the laboratory. This software, due to the large size of the potential market is often of a very high standard and good value for money. By comparison, scientific programs are significantly more expensive and of rather variable quality. One reason for this is that the total market for a scientific package is substantially smaller than for business software making it difficult to recoup the development costs while selling the software at prices acceptable to the average scientific user. The specialized nature of many types of applications, such as single channel analysis, also requires the developer to have a detailed knowledge of the subject. Most of the software that is commercially available has been originally developed in an academic laboratory. Nevertheless, if a particular program fits one's needs it is probably better to buy than to develop one's own. When considering such a purchase it is useful to inquire whether the software is used in other similar laboratories and to obtain their opinions on the software. If the cost of the software is substantial, a prospective purchaser would be also advised to obtain an adequate demonstration of the software working in the their own laboratory before committing themselves.

6.5.1 *ASYST*

The program known as ASYST (from the Macmillan Software Company) is one of the few attempts at providing a truly versatile signal acquisition and analysis package. It is a command driven system capable of supporting laboratory interfaces from a number of manufacturers including, Data Translation, Cambridge Electronic Design, and Tecmar. It can display digitized data files in a number of formats and provides operations ranging from simple addition and multiplication of arrays of data to FFTs, curve fitting and waveform smoothing. It also has a multi-tasking facility which allows display and arithmetic operations to be performed on stored arrays while others are being collected. ASYST commands can be collected together and executed from within batch files speeding up routine sequences of operations. In fact complete programs can be written in this way. Since the ASYST commands are directly suited to signal acquisition and analysis it is much simpler and quicker to develop applications in this environment than with BASIC. ASYST however, because it is in essence an interpreter, is substantially slower in operation than a custom built FORTRAN program. In many cases this may not be too high a price to pay for the short time taken to develop a program but might be crucial in other computationally intensive applications such as single channel analysis.

6.5.2 CED electrophysiology software

Cambridge Electronic Design take another approach and attempt to provide highly customized applications programs for specific types of analysis (not only electrophysiology). At the time of writing, they have announced several packages within a modular system of software for electrophysiological analysis. They have yet to be tried and tested in daily use in a large number of laboratories so it is difficult to comment on their ultimate value. However, the company appears to have a fairly long term commitment to this market making their products certainly worthy of consideration. The software is written in Pascal and makes use of the METAWINDOWS graphics package.

6.5.3 PCLAMP

PCLAMP is a suite of data acquisition and analysis programs for voltage-clamp and patch-clamp experiments, originally developed at Cal. Tech. (13), but now marketed by Axon Instruments Inc. (PCLAMP, Axon Instruments Inc.). The software has been constantly improved, currently reaching version 4. It is written using the Microsoft QuickBASIC 2.0 compiler, plus some assembler code routines, and provides a variety of features similar to the software described in this chapter. The Tecmar Labmaster is required as the laboratory interface, providing ADC sampling rates between $33-70$ kHz depending on the computer. The lack of a DMA facility on the Tecmar board is, however, a distinct limitation preventing continuous ADC sampling to a disc file. On the other hand, the fact that the complete source code is provided in a simple and accessible language such as BASIC allows the user to customize this program for their own purposes.

7. ACKNOWLEDGEMENTS

The author would like to thank Drs Alisdair Gibb, Jerry Lambert, Ian G.Marshall, John Peters and Chris Prior for the use of experimental data. This work was supported by the Wellcome Trust and the Organon Scientific Development Fund.

The software described in this chapter is available from the author, free of charge to academic and non-commercial users. A CED 1401, DT2801A, or DT2821 interface is required for its operation.

8. REFERENCES

1. Hille,B. (1984) *Ionic Channels of Excitable Membranes*. Sinauer Associates Inc., MA, USA.
2. Sakmann,B. and Neher,E. (1983) *Single Channel Recording*. Plenum Press, New York, USA.
3. Eggebrecht,L.C. (1983) *Interfacing the IBM Personal Computer*. The Waite Group, USA.
4. Horowitz,P. and Hill,W. (1980) *The Art of Electronics*. Cambridge University Press, Cambridge, UK.
5. Press,W.H., Flannery,B.P., Teukolsky,S.A and Vettering,W.T. (1986) *Numerical Recipes. The Art of Scientific Computing*. Cambridge University Press, Cambridge, UK.
6. Colquhoun,D. and Sakmann,B. (1985) *J. Physiol.*, **369**, 501.
7. Bormann,J., Hamill,O.P. and Sakmann,B. (1987) *J. Physiol.*, **385**, 243.
8. Bendat,J.S and Piersol,A.G. (1971) *Random Data: Analysis and Measurement Procedures*. John Wiley and Sons Inc., New York, USA.
9. Griffiths,P. and Hill,I.D. (eds) (1985) *Applied Statistics Algorithms*. Ellis Horwood Ltd, Chichester, UK.
10. Eisenberg,R.S., Frank,M. and Stevens,C.F. (1984) *Membranes, Channels and Noise*. Plenum Press, New York, USA.
11. DeFelice,L.J. (1981) *Introduction to Membrane Noise*. Plenum Press, New York, USA.
12. Lamb,T.D. (1985) *J. Neurosci. Methods*, **15**, 1.
13. Kegel,D.R., Wolf,B.D., Sheridan,R.E. and Lester,H.A. (1985) *J. Neurosci. Methods*, **12**, 317.

CHAPTER 3

Anatomical measurement and analysis

JOSEPH J. CAPOWSKI

1. INTRODUCTION: WHY APPLY COMPUTERS TO NEUROANATOMY?

Neuroanatomists have classically had primarily a qualitative training, looking at anatomical structures, both large and microscopic and describing what they see. In the last decade, however, this has begun to change. The introduction of the personal computer and its auxiliary devices into the laboratory have made it reasonable for the anatomist to enter anatomical information into the computer and generate numeric information about the structures. Thus a structure may be characterized mathematically and one population of structures compared to another.

1.1 The combination of human and computer

A human being is a very good pattern recognizer. He can glance at an image and in a very short time, extract a lot of information from it. He can also easily and selectively ignore information about which he has no concern. Should he choose to do so, he may even look at the overall form of an image, ignoring details, for example to observe the shape of a tree while paying no attention to its leaves and twigs. However, a person is a very poor book-keeper. When adding lists of numbers, he works slowly and makes errors.

Contrast him to a computer. A computer is an excellent book-keeper, which, once programmed, can deal quickly and accurately with long lists of numbers. The computer cannot easily recognize patterns, however. It indeed can capture an image into its memory and ask questions of the image such as 'What is the grey level of a small picture element (pixel) at coordinate 330,195 in the picture?'. The computer program can, with reasonable effort, deal with very small parts of a picture. However it is very difficult to write a computer program to look at an entire picture and to recognize patterns. This would require a program approaching the level of sophistication of the human visual system. Furthermore the space and computing power required are usually not within the budget of a neuroscience researcher.

However, the combination of a human working with a computer is most effective. The person can do what he does best, pattern recognition, and avoid what he does worst, book-keeping. The computer, with a well-written program becomes his fast and accurate book-keeper, making no attempt at pattern recognition.

1.2 A semiautomatic system for collecting data

It is possible to assemble a system for the collection of anatomical data using the principles just illustrated. An anatomist looks at an image and using his visual skills,

Anatomical measurement and analysis

Figure 1. A researcher outlining a structure with a data tablet. The x,y coordinate of the cross-hairs, inside the circle above and to the left of her hand, can be sensed by the computer. The computer program can also sense the status of the four pushbuttons.

passes a cursor over the image, perhaps tracing an outline. The computer keeps track of the position of the cursor and records a series of these positions in its memory. This list of computer-stored coordinates now forms a model of the just-traced outline. The computer may display the outline on a cathode ray tube (CRT) to provide feedback that the tracing process is progressing well and may calculate, for example, the area enclosed by the outline. These human−computer systems are called 'semiautomatic' because they automate the data entering process to a degree. The bulk of this chapter is used to describe some semiautomatic computer systems which are in common use today in neuroanatomical laboratories.

The semiautomatic system should be contrasted to a fully automatic system in which the computer, rather than the human, looks at the image. Here a sophisticated program attempts to perform some of the functions of the human visual systems. These form excellent computer science research projects, but are not yet effective biological research tools in common use in the laboratory.

2. SEMIAUTOMATIC TRACING WITH A CURSOR

2.1 Techniques of tablet tracing

Frequently, an anatomist wants to extract numerical information from an image. A data tablet is a common computer input device designed for this task. *Figure 1* shows a researcher tracing a photographed anatomical image using the tablet's cursor.

An acoustic data tablet is the easiest type to describe. A flat surface is provided on

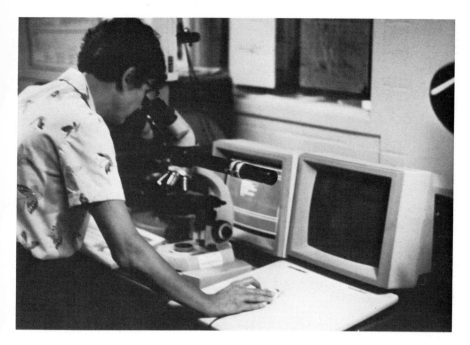

Figure 2. A researcher using a microscope with a drawing tube or camera lucida and a data tablet. The drawing tube's optical input port is looking down at the data tablet.

which to place a photograph. The user holds a stylus which looks very much like a ball point pen. At frequent intervals, say 10 times per second, the stylus generates a spark and an accompanying 'click' sound. Simultaneously the computer inside the data tablet starts two numbers counting at a fixed rate. The sound wave radiates in a circle from the stylus like ripples on the surface of a pond caused by dropping a stone into the water. At the top and right edges of the flat surface are two strip microphones. When the sound wave reaches the microphone on the right edge of the surface, the microphone senses the sound and signals the data tablet computer to stop counting one of the counters. The value now contained in the counter is proportional to the time required for the sound wave to reach the microphone and thus to the distance from the stylus to the right edge of the tablet. Similarly the top microphone senses the sound wave that reaches it and signals the computer to turn off its counter. Thus the tablet can calculate the current x,y position of the cursor. These values are passed to the main computer, so that its program always knows where the user has positioned his hand. Other tablets use the same concept of a radiating wave, but in a different medium. The wave may be electromagnetic propagating in a wire mesh inside the tablet rather than an acoustic wave propagating in the air.

A data tablet stylus usually takes one of two forms. The first type is like a ball point pen with a switch in the pen tip to sense if the user has put the pen down on the tablet surface. The computer can sense the position of the stylus and whether the pen is up or down. The second type of cursor (shown in *Figure 1*) consists of cross-hairs and several pushbuttons. The computer can sense the position of the cross-hairs as the user

Anatomical measurement and analysis

Figure 3. The view in the microscope eyepieces when using the drawing tube with a data tablet. The cursor's cross-hairs can be seen overlaid on the tissue image. The tissue is cat neocortex with numerous Golgi-stained cells shown. Courtesy of Richard J.Weinberg, UNC Department of Anatomy.

traces and can sense which button has been depressed. He may use the pushbuttons to signal the computer software to perform any number of functions.

Note that a tablet has an absolute coordinate system, ranging from say 0,0 at the lower left to 1000,1000 at the upper right. Even if the stylus is lifted and replaced, the computer always reads the same coordinate if the stylus is placed at the same location. This is in contrast to the mouse, as will be explained later.

The anatomist traces by moving the tablet cursor over the image. He must first, however, get the image onto the tablet. Several ways are commonly used for this. The first is simply to place a photograph of the image on the tablet as shown in *Figure 1*.

The second method of placing the image onto the tablet is by using a camera lucida, or drawing tube, as shown in *Figure 2*. A drawing tube is a microscope attachment for mixing an external image with the microscopic image and presenting the combined image to the viewer in the microscope eyepieces. The external image is, in this case, the tablet cursor and the researcher may move this over the specimen. Frequently a little light bulb or light-emitting diode (LED) is placed at the tablet cross-hairs in order to make it easier to see in the microscope eyepieces. The microscope view is shown in *Figure 3*.

A third method of placing the image on the data tablet is to project the image onto the tablet using a slide projector or a projecting microscope. Transparent and translucent data tablets are manufactured in order to allow the projection to be done from the rear of the tablet.

Figure 4. The researcher outlining a tissue section with a mouse. The just-traced outline is displayed on the screen so that she can see that the tracing has proceeded correctly. She has just moved the mouse down on its pad and the cursor (a small cross) has moved down the TV monitor from the just-traced outline. The tissue is the dorsal horn of rat spinal cord. Courtesy of Elizabeth Bullitt, UNC Department of Neurosurgery.

Once the image is placed on the tablet, the user may locate or may outline structures. If his research interest is only the number of structures or perhaps additionally, their spatial distribution, then he simply needs to position the cursor at each structure and press a button. The computer software will record the coordinate of the structure. If he is interested in more sophisticated measurements of each structure, he may want to trace or outline it. Then the computer records a series of coordinates which define the shape of each structure.

2.2 Feedback is important

Whatever system is employed for presenting the image and passing a cursor over it, it is mandatory that feedback be provided to the user that the computer is recording the data correctly. The best possible feedback that tracing is proceeding well is provided by a computer-generated outline overlaid on top of the original image. *Figure 4* shows a researcher tracing a structure with a tablet cursor. As he enters the outline, he may look on the screen to see the computer-stored data. He can easily confirm that the tracing is proceeding nicely. Should he make a mistake, he can delete the erroneous portion of the trace and continue as if no mistake had been made.

A mouse is a computer input device with similarities to the data tablet (Summagraphics). It consists of a small, hand held box with two friction wheels on its bottom. The axes of the two wheels are at right angles to each other. The user rolls

99

Anatomical measurement and analysis

Figure 5. A Hewlett-Packard model 7475A felt tip pen plotter. The computer drives the pen in the right−left direction while moving the paper in the near−far direction to plot a structure on the paper. Various colour pens may also be selected by the computer.

the mouse on the table and the computer can sense the amount that the wheels have rotated and compute the distance that the user has moved the mouse in both directions. It is possible to use a mouse to move a cursor on a screen, but note that the mouse has a relative coordinate system; that is, the mouse can be lifted and placed anywhere without the computer knowing that it has been moved.

3. ELEMENTARY TWO-DIMENSIONAL TABLET MEASUREMENTS

The major advantage of entering data into a computer memory is that the memory is tractable. It is possible, with reasonable effort, to write computer programs which will present the data visually, overcoming the constraints of the light microscope. It is also possible to write programs to characterize the data mathematically and to compare one population to another. This section describes some common two-dimensional displays, measurements, and calculations which are used to quantify neuroanatomical structures.

3.1 Marking the locations of cells and other structures

Figure 3 shows a low power view of numerous neurons in one section of a region of the brain. Not all the cells are in focus, and most are not even shown because they are in different tissue sections. The researcher may pass a cursor over the microscope image, outlining the tissue edge and the soma of each cell, and indicating that each cell is of a certain type. For this example, cells might be categorized by the number

J.J.Capowski

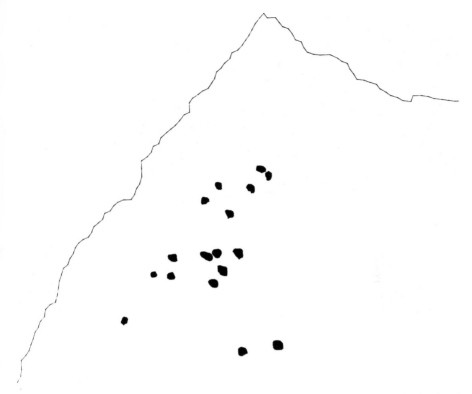

Figure 6. A plot of the somas of the neurons in the tissue of *Figure 3*. The outline of the tissue is also plotted.

of dendrites which emanate from their somas. This is done for all the neurons in the section, over the entire focus range, and for several tissue sections.

Some device must be used to make a copy on paper (in computer terminology, a 'hard copy') of the result. This task may be performed by a felt tip pen plotter (Hewlett-Packard), shown in *Figure 5*. The result from the plotter is presented in *Figure 6*. This plot is a map of all the cells in the tissue volume.

3.2 Counting features

When the structures have been traced into the computer memory using the data tablet, it is a reasonable task to write a computer program to generate mathematical summaries of the data. Counts of features are the lowest level mathematical summaries. So for the data of *Figures 3* and *6*, we could state that there are 17 neurons in this region of the brain.

Counts can also be made from data such as shown in *Figure 7*. Here individual structures which are approximately circular are traced into the computer using a data tablet. A category number may be entered during tracing, so that the structures may be classified by any visual criteria which the user chooses. When the tracing is finished, the user may request a list of the number of structures in each category.

Anatomical measurement and analysis

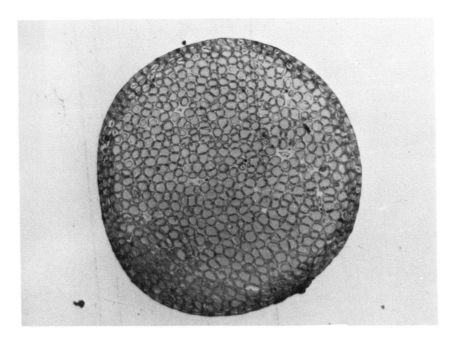

Figure 7. The cross-section of a motor neuron nerve root from the cat. Myelin sheaths can be seen surrounding each axon in the root. Courtesy of Dominic Sinicropi, Genetek Corp.

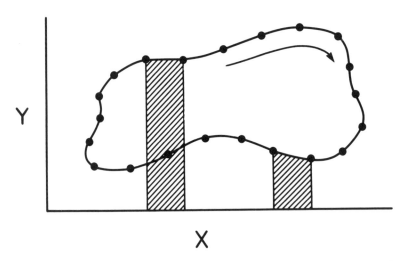

Figure 8. Using the trapezoidal rule to calculate the area of an irregular structure. The area of each trapezoid is calculated and summed. If tracing is clockwise, then the calculated area will be positive; if counterclockwise, the area will be negative and must be negated. More details are given in the text.

3.3 Calculation of two-dimensional structural parameters

Frequently, counts of structures are not sufficient to describe the structures. It is also necessary to present some mathematical parameters which describe the shape of each

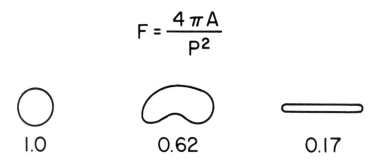

Figure 9. Values of the form factor parameter for various outline shapes.

structure. In the past two decades, several shape parameters have become standard in the neuroanatomical field. These are described below.

Straight line distance may be calculated from any coordinate in a structure to any other coordinate. This is done with the Pythagorian theorem from any two coordinates which are read from the data tablet cursor.

The calculation of the cross-sectional area of an irregular outline is more complicated. As shown in *Figure 8*, the trapezoidal rule may be used to calculate area. Here the area is divided into columns and the area of each column (its width times its mean height) is summed to form the area of the outline. The coordinates read from the data tablet may be used conveniently to form the upper corners of the columns and an arbitrary x axis is used to form the bottom of the columns. If the outline is traced clockwise, the area will be calculated correctly, but if it is traced counterclockwise, the computed area will have the right magnitude, but will be negative. It is only necessary then to make it positive. This is an easy computer program to write and it gives reasonably accurate results.

Once area has been calculated, it is easy to calculate the mean diameter and the perimeter of the outline. If we assume that the outline is a circle, then these values are:

$$\text{Mean diameter} = 2 \times \sqrt{(\text{Area}/\pi)}$$
$$\text{Perimeter} = 2 \times \sqrt{(\pi \times \text{Area})}$$

Another frequently used descriptive parameter of an outline is its 'form factor'. This is a measure of roundness, calculated by the formula:

$$\text{Form factor} = 4 \times \pi \times \text{Area}/(\text{Perimeter})^2$$

Figure 9 shows the form factors for several outline shapes.

The parameters described above are measures of size and shape, but do not indicate the directional orientation of a structure. To indicate orientation, a major axis calculation may be performed, as shown in *Figure 10*. Here the computer program finds the pair of coordinates in the structure which are farthest apart. It then draws a straight line through the pair. The angle of this line from the horizontal indicates the structure's orientation.

Anatomical measurement and analysis

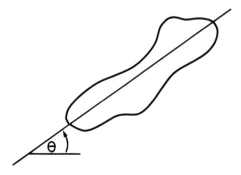

Figure 10. The major axis of a structure. The angle theta measures the orientation of the structure from the horizontal.

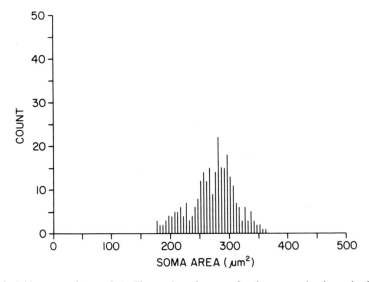

Figure 11. A histogram of soma sizes. The number of somas of each cross-sectional area is plotted.

The centre of gravity of an outline may also be calculated. In engineering, a structure's centre of gravity is the coordinate at which all the structure's mass may be concentrated and the structure's actions upon its environment will not be changed. The term is used loosely in its application to neuroanatomy to mean the centre of an irregularly-shaped structure. It is calculated by first assuming an arbitrary x,y coordinate. Then individual elements of area times the distance from the arbitrary coordinate are summed. The result is divided by the sum of all the areas. This yields a coordinate which represents the middle of the structure.

4. POPULATION CALCULATIONS FROM THESE PARAMETERS

All of the above numeric values may be calculated for individual outlines. Frequently however, a neuroanatomist wishes to compare one group of outlines to another group.

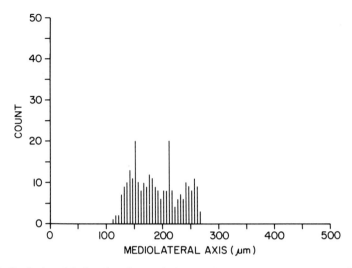

Figure 12. A distribution of the location of somas in the mediolateral direction to show how the somas are distributed spatially.

To accomplish this, the numeric values for each structure in a population should be calculated and a summary of the population presented. There are literally thousands of ways to compare populations of structures, so I shall only present some illustrative examples.

4.1 Histograms

Simple histograms may be generated for any numeric parameter, for example, area of outlines. 'Bins' of area are defined, and each one has a range, say, $0-100$ μm^2, $100-200$ μm^2, etc. Then the computer program calculates the area of each outline and determines into which bin the outline falls. The number of outlines which fall into each bin is plotted versus the size of the bin. *Figure 11* shows such a histogram.

4.2 Distributions

Frequently, one wants to show how some numeric parameter is distributed within the tissue. For example, as is shown in *Figure 12*, the parameter 'cell location' is plotted against some anatomical axis. Any parameter: location, size, area, number of synaptic contacts, etc. may be distributed in such a fashion. Distributions may be plotted against an anatomical axis, or against a more complex parameter, such as radial distance from the soma of a neuron, or thickness of a neuronal process.

4.3 Density measurements

The term 'density' used here refers to the number of occurrences of some feature per unit length, area, or volume. A common example in neuroanatomy is to calculate dendritic spine density. The number of spines along a dendrite is counted and the count is divided by the total length of the dendrites to yield the density. Another frequent

Anatomical measurement and analysis

density is the number of occurrences of a structure per unit volume. So for example, if neurons in one region of the brain are stained and a different region of the brain is examined, the number of stained cells counted in the second region indicates the amount of projection of the original cells into the second region. By expressing the count as a volume density instead of as an absolute count, it is possible to normalize for the size of varying brains and brain regions.

This use of the term 'density' should be contrasted to another use, video 'density', which is described later in the chapter. There, density refers to the grey level or amount of brightness of a particular point in a picture.

5. THREE-DIMENSIONAL COMPUTER GRAPHICS

The next sections of this chapter refer to the display and analysis of three-dimensional structures. In order to describe the display of such structures, it is worthwhile to present a brief introduction to the problems of generating a three-dimensional image on a two-dimensional surface, either a computer CRT or a sheet of paper.

Computer graphics hardware which is used to draw a picture from data stored in computer memory, may be divided into two general categories: raster and vector. In their most frequent implementations, both types utilize a CRT on which to present the image. The CRT consists of a glass bottle whose interior has been evacuated. An electron beam is emitted from the rear of the bottle and is attracted toward the front surface of the bottle. It is focused sharply so that the electrons reach the front surface in a narrow beam. The interior front surface of the bottle is coated with phosphor which has the property that when an electron strikes it, it emits a photon of light. Thus a bright dot is presented on the front surface of the bottle at the point where the electron beam strikes the phosphor coating. The difference between a raster and a vector system is determined by how the electron beam is deflected over the front surface of the CRT.

In a raster display system, the most common example of which is television, the electron beam is swept across the entire face of the CRT in a fixed pattern of horizontal lines called a raster, regardless of what is drawn on the CRT. In order to present information on the screen, the intensity of the electron beam is modulated. When driven by a computer, each line is divided into a number of points called picture elements, or 'pixels'. There are usually about 500 pixels in a line and about 500 lines in a raster, so that a picture's resolution may be on the order of 500 by 500 elements. The electron beam traverses its entire raster every 33 msec. For each point, the computer must specify a light intensity value and/or a colour. This is a lot of information for a single image, usually about 250 000 pixels at one byte per pixel. The raster display is a powerful presentation technique which allows a computer to present an image composed of text or structures with realistic surfaces shown. Raster images are presented in *Figures 13 and 15*.

The alternative is a vector display system, the most common example of which is the laboratory oscilloscope. The electron beam is deflected only to those places on the CRT's face where lines are to be drawn. Typically, the electron beam may be deflected with much greater precision than with a raster system. The time required to draw an

Articulated Figure Positioning by Multiple Constraints

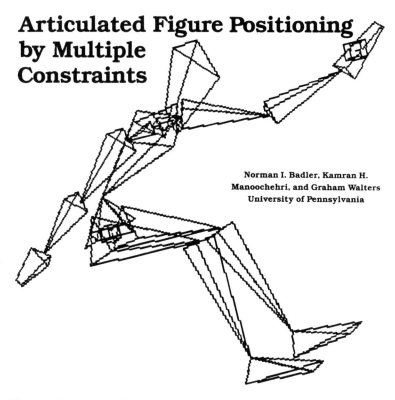

Norman I. Badler, Kamran H. Manoochehri, and Graham Walters
University of Pennsylvania

Figure 13. A wire frame structure drawn on a moderate resolution raster screen. The resolution of the screen can be seen easily as a staircase effect on the straight lines. The worst case shows when the lines are nearly horizontal. Reproduced from *IEEE Computer Graphics and Applications*, June 1987, p. 28, with permission.

image depends upon the complexity of the image, but is generally much shorter than the time required to draw a raster image. An image is made up of a series of straight lines and the computer only need store the coordinate of each line endpoint, hundreds or at most, several thousand coordinates. Text and 'wire frame' structures (structures only whose outlines are shown, as if they were made of wire) can be drawn easily with a vector system. A vector image is presented in *Figure 14*.

Both of these systems face the same challenge: how to present the illusion of a three-dimensional image on a two-dimensional screen? In order to help create the illusion of depth, several techniques have been developed in the computer graphics field. These techniques are called 'depth cues'.

A common depth cue for looking at a three-dimensional structure is smooth rotation. If the structure is slowly and smoothly rotated on the screen, the visual system of the viewer will fool him into thinking he is seeing a three-dimensional object. Another depth cue is perspective division. Since we are used to seeing distant objects smaller, if objects, or parts of them, are presented as smaller, we will perceive them as farther away. This technique is much more useful however, for looking at three-

Anatomical measurement and analysis

DMP-51 handles bigger (ANSI C/D) paper sizes, which made it an appropriate tool for evaluating the quality of larger output.

Once Prodesign II is configured for your hardware, type PD to start the program. An X-shaped cursor appears in the middle of the screen and a status line at the top. The Ins key (or an appropriately configured mouse button) is used to pick points and the Esc key deletes points, lines, or arcs. Press G to toggle through various grid styles.

To draw lines, press Ins, move the cursor, press Ins, then press V (for vector) and a line appears on the screen. To draw free-form curves, press Ins, move the cursor, repeat to specify any number of points, then press C to fit a curve to the points that were selected. Press Esc and the last-drawn entity disappears. To copy, move, or rotate objects, you define a section and then issue the appropriate command.

These few examples should give you a pretty good idea of how the program works. The logical command structure makes Prodesign II easy to use. To make things even easier, the manufacturer includes a command template that can be taped to a digitizing tablet. With

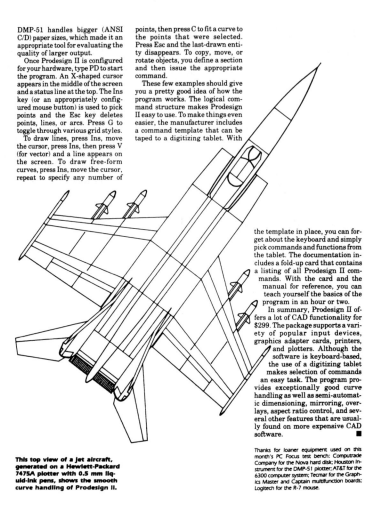

the template in place, you can forget about the keyboard and simply pick commands and functions from the tablet. The documentation includes a fold-up card that contains a listing of all Prodesign II commands. With the card and the manual for reference, you can teach yourself the basics of the program in an hour or two.

In summary, Prodesign II offers a lot of CAD functionality for $299. The package supports a variety of popular input devices, graphics adapter cards, printers, and plotters. Although the software is keyboard-based, the use of a digitizing tablet makes selection of commands an easy task. The program provides exceptionally good curve handling as well as semi-automatic dimensioning, mirroring, overlays, aspect ratio control, and several other features that are usually found on more expensive CAD software. ■

Thanks for loaner equipment used on this month's PC Focus test bench: Computrade Company for the Nova hard disk; Houston Instrument for the DMP-51 plotter; AT&T for the 6300 computer system; Tecmar for the Graphics Master and Captain multifunction boards; Logitech for the R-7 mouse.

This top view of a jet aircraft, generated on a Hewlett-Packard 7475A plotter with 0.5 mm liquid-ink pens, shows the smooth curve handling of Prodesign II.

Figure 14. A wire frame structure drawn on a high resolution vector plotter. The straight lines show no staircase effects. Reproduced from *Computer Graphics World*, December 1985, p. 72, with permission.

dimensional environments than for looking at individual structures.

Intensity depth cueing is sometimes used. Since we are used to seeing distant objects dimmer, then if objects, or parts of them are presented dimmer, we will perceive them as farther away. This technique is also only of marginal use in looking at structures.

Hidden line or hidden surface removal is a most valuable depth cue for looking at structures. We are used to seeing a near surface of a structure block our view of a distant surface. Hence if the computer can calculate the edges of each surface that block those behind it, it may simply not draw, or 'hide' the hidden surface. *Figure 15* shows a hidden-surfaced structure drawn on a raster system.

A final depth cue is the presentation of stereo pairs. Here two images of the same structure are presented, with one rotated approximately six degrees about the structure's

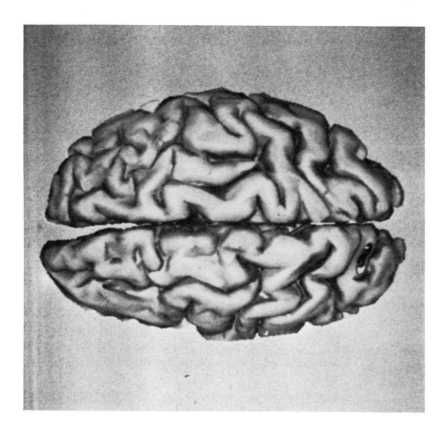

Figure 15. A 'surfaced' representation of the cerebral cortex drawn on a high resolution raster screen. Reproduced from *Computer Graphics*, February 1980, with permission.

y axis from the other. Some optical technique (e.g. coloured or polarized filters, prisms, lenses) is used to direct one image to the viewer's left eye and the other image to the viewer's right eye. The human visual system merges these two two-dimensional images into a single three-dimensional image. Furthermore, by changing the relative rotation angle, it is possible to exaggerate the effect to emphasize the depth of the structure. *Figure 16* shows a plot of stereo pairs of a neuron. Some people may, with no optical aid, be able to see the depth in the structure by relaxing their eyes and merging the images.

The two types of graphics systems each have advantages and disadvantages for displaying three-dimensional structures with these depth cues. The raster system allows a more realistic presentation of a structure with its surfaces filled, such as *Figure 15*. The structure can be presented in colour and with subtle shading, reflections, and even shadows. However, the massive amount of data required to present the structure demands either a more powerful computer or a longer length of time to build the display. Once built, it is not possible to rotate the display smoothly so to enhance its three-

Anatomical measurement and analysis

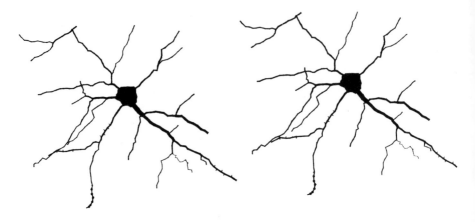

Figure 16. A plot of stereo pairs of a Golgi-stained neuron, plotted by the UNC neuron tracing system.

dimensionality. The vector system may be used to present a rotating presentation, but only of a wire frame model.

The spatial resolution of a raster system is generally poorer than that of a vector system. Compare, for example, *Figure 13* with *Figure 14*. On the other hand, since the raster system takes advantage of the mass consumer market of television, the hardware to present a raster image is generally less expensive.

6. THREE-DIMENSIONAL SERIAL SECTION RECONSTRUCTION

The interior of a three-dimensional structure such as a brain is often studied by cutting it into a series of flat slices or sections. While this does allow one to study the interior of the brain, it destroys its intact three-dimensional features. With the technique of computer-assisted serial section reconstruction, it is possible to reassemble or 'reconstruct' the brain inside a computer memory in all three dimensions and then generate displays and statistical summaries of the reconstructed brain.

To do a serial section reconstruction, individual sections (or photographs of them) are placed on the data tablet. Each section is traced with the data tablet cursor, and the computer records a series of x,y coordinates which define that two-dimensional section. A z coordinate, representing that section's depth into the tissue is entered once for each section and is appended by the computer to each x,y coordinate, resulting in a series of x,y,z coordinates for each traced outline. When a series of sections has been traced, the coordinates of the three-dimensional structure are now recorded, and the structure, or more properly, a mathematical model of it, has been reconstructed in the computer memory. This is the mathematical equivalent of stacking the tissue sections. Once the model is in computer memory, it is a reasonable task, using the graphics techniques described above, to build a three-dimensional display of the structure on the CRT. A bat larynx reconstruction is shown in *Figure 17*. The display can be smoothly rotated or hidden lines can be removed to help the viewer understand its three-dimensional structure. Commercial software is available (Eutectic Electronics Inc.).

J.J. Capowski

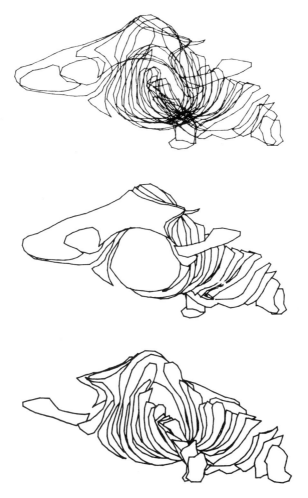

Figure 17. A three-dimensional serial section reconstruction of a bat larynx. **Top**: the stacked sections are rotated about the *y* axis to an oblique angle, but with all lines drawn it is not possible to tell which is the front and which is the rear. **Middle** and **Bottom**: Front and rear views of the structure with hidden lines removed to clear up the ambiguity. Reconstruction done by James Kobler, Department of Anatomy, Harvard University. Reproduced from *J. Neurosci. Methods* (1985) **13**, p. 146, with permission.

6.1 Section alignment

A classic problem in serial section reconstruction is that of aligning one tissue section to another. For an irregularly-shaped structure such as the brain, when the tissue is sectioned, the sequential sections may not be of the same size and shape. However some alignment must be performed in order to ensure that the resulting structure is a faithful model of the original, unsectioned one. Two regular landmarks which extend through the structure perpendicular to the plane of sectioning may be used to help align one section to another. Two adjacent sections must be overlaid on the tablet so that the landmarks coincide on the tablet. Then rotational and translational alignment of the two sections is assured. The overlay may be performed optically or with computer aid by displaying one section on the CRT and pointing at the landmarks with the tablet cursor while moving the second section until landmark locations of the two sections coincide. If no regular landmarks are present, it may be possible to introduce them before sectioning by, for example, piercing the tissue with a pin. If this is not possible, it may be necessary to simply guess at the alignment by overlaying two sections and matching them up by eye as well as possible.

6.2 Three-dimensional measurements

Three-dimensional measurements may also be performed on the reconstructed structure with reasonable effort. A distance can be measured by moving a cursor to one point on the structure (the data tablet cursor can be moved directly in the x and y directions and the z value can be entered from the keyboard, or the computer can mathematically determine the x,y,z coordinate of the point in the structure nearest the cursor) and then to another point. The computer can then calculate the distance between the two points using the Pythagorian theorem.

Volume measurements may be easily calculated. The cross-sectional area of corresponding two-dimensional outlines in adjacent tissue sections may be calculated as described in *Figure 8*. By averaging these two areas and multiplying the result by the thickness of the tissue between their two sections, the volume of the structure surrounded by the areas and extending between the sections may be calculated. This calculation may be extended to a long series of adjacent sections.

A surface area calculation is also frequently done. Here the perimeter of the outlines is calculated and the mean perimeter is multiplied times the slab thickness. The result is the area of the surface band formed by the two outlines. Any of these just-calculated parameters may, of course be combined into population measures or distributions, such as shown in *Figures 11* and *12*.

7. NEURON RECONSTRUCTION

A more complicated three-dimensional reconstruction procedure is required at light microscopic magnifications for neurons found in thick tissue sections. As *Figure 19* shows, two characteristics of neurons make the problem more difficult. First, the dendrites of the neurons meander in three dimensions within a slab of tissue which is perhaps 80 μm thick. Therefore it is necessary to capture a series of x,y,z coordinates where the z value will vary from point to point along a dendrite within one tissue section. Second, the dendrites branch repeatedly in tree-like fashion, so that any data collection program which is written must accommodate tree structures, not simply outlines and dots as described in the data tablet tracing section.

7.1 Computer hardware for tracing neurons

For about 20 years, teams of neuroscientists and engineers in several university laboratories have been working to design a computer system to quantify the branching patterns of axons and dendrites. With the recent lower prices of small computers and graphics display systems, perhaps forty systems of various complexity have been built.

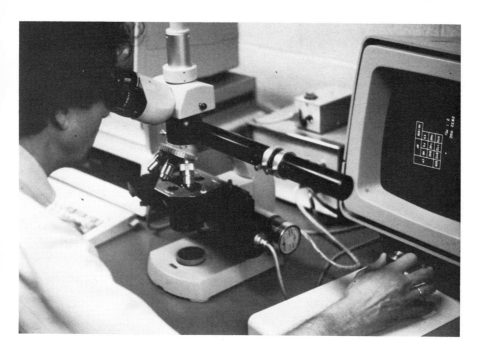

Figure 18. A researcher tracing a neuron with the UNC neuron tracing system. On the large CRT is a computer-generated display. The drawing tube mixes this display with the neuronal image and presents the combined image in the microscope eyepieces (*Figure 19*). As the joystick is controlled, stepping motors drive the microscope stage and focus axis, allowing the dendrites to be followed. Buttons on the computer keypad are pressed with the operator's left hand to signal the computer to record points along the dendrite.

In this section the neuron tracing system at the University of North Carolina will be described in order to illustrate the principles of neuron tracing, display and analysis.

The important hardware in the UNC neuron tracing system is shown in *Figure 18*. A light microscope (Zeiss) has been fitted with stepping motors to drive each of the three axes (x,y, and focus) with a resolution of 0.5 μm. The motors are driven by an IBM AT computer with 512 K of memory. The researcher controls the position of a joystick; its output is read by the computer through an analogue-to-digital converter board. The joystick is used to control the motion of the microscope stage in the x and y directions. As the researcher deflects the joystick, the microscope stage tracks his motion. The joystick is 'three-dimensional'; that is, it has a potentiometer on its top. This potentiometer is read by the computer and is used to control the focus axis of the microscope. An additional potentiometer is located to the right of the joystick which the researcher may simultaneously control with the little finger of his hand. The computer senses the position of this potentiometer also. A vector computer graphics system is attached to the computer and presents its output on a large CRT just behind the joystick. The image on the CRT is mixed with the microscope image by a camera lucida and the combined image is presented to the viewer in the microscope eyepieces as shown in *Figure 19*. To the user's left is the computer's typewriter-like keyboard and its keypad, the separate cluster of keys on its right end.

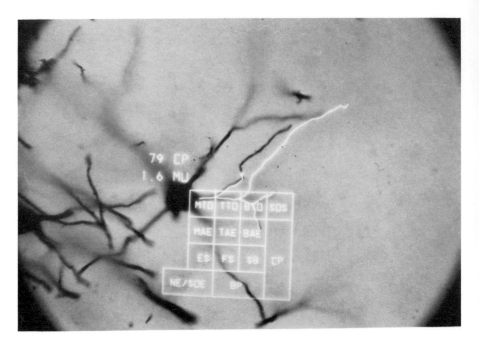

Figure 19. The view in the microscope eyepieces during the tracing of a dendrite. The tissue is a high magnification image of *Figure 3* and contains a Golgi-stained neuron in the cat neocortex. The tissue is is 80 μm thick and the dendrites can be seen meandering in all three directions, especially in and out of focus. A computer display appears overlaid on top on the neuron image. The large grid is a map of the computer keypad, indicating which buttons to press to record various types of points. A circular cursor which is passed over the dendrite is shown just above the grid. The user sets its diameter to that of the dendrite so that the computer may record the thickness of the dendrite. The computer presents the just-traced dendrite so that the viewer can verify that tracing is proceeding correctly. The number '79' is the number of points which the computer has captured; 'CP' indicates that the most recently entered point is a 'continuation point', and '1.6 MU' indicates the current cursor diameter.

7.2 The neuron tracing procedure

When the researcher looks into the microscope, the image shown in *Figure 19* is seen. The following procedure should then be carried out.

(i) Move the joystick in $x,y,$ and focus so to follow the dendrites by passing them under the cursor.

(ii) Adjust the potentiometer to the right of the joystick to maintain the diameter of the cursor equal to that of the dendrite.

(iii) At spatial intervals along the dendrites type a key on the computer keypad. The computer then records the point, its type, its current x,y,z coordinate and its thickness. A series of these points, when recorded in computer memory, form a mathematical reconstruction of the tree. Note that, unlike the case in serial section reconstruction, each point has a specific z coordinate, taken from the focus axis of the microscope.

(iv) Whenever a branchpoint is encountered press the 'BP' button. The computer will remember the branchpoint.

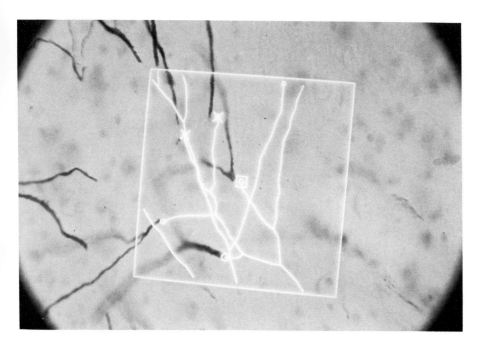

Figure 20. The view in the microscope during the 'merging while tracing' procedure. A computer display of the already-traced trees from one tissue section is generated on top of the next tissue section. Circles in the display indicate endpoints to which trees should be attached. The tissue may be rotated and translated so that the start of its tree pieces coincide with the circled endpoints. Crosses indicate endpoints to which traced trees have already been attached. The square identifies the endpoint to which the next tree traced will be attached.

(v)   Trace either exiting path from the branchpoint, and when an ending is reached, press the 'NE' button to indicate a natural end. The computer will then move the microscope stage and focus back to the branchpoint in order to prompt the user to trace the remaining path from the branchpoint. This is how repeatedly branching structures are recorded.

7.3 Merging of multiple section dendrites

When tissue is sectioned, neuron trees are frequently cut into pieces which must be somehow re-attached in order to generate a realistic reconstruction. There are two methods for performing the merges: 'merging during tracing' and 'merging after tracing'. *Figure 20* illustrates the former technique.

The computer may generate an image of the already-traced trees of a previous section overlaid on the biological image of the current section. The researcher may move the current section in the microscope until its tree starting points match the stumps of the trees in the previous section. The computer software allows identification of one of the stumps. Tracing of the new tree is then begun and the computer is instructed to attach the new tree to the identified stump. The computer translates all the coordinates of the new tree so as to place it exactly at the stump and to continue the old tree.

Anatomical measurement and analysis

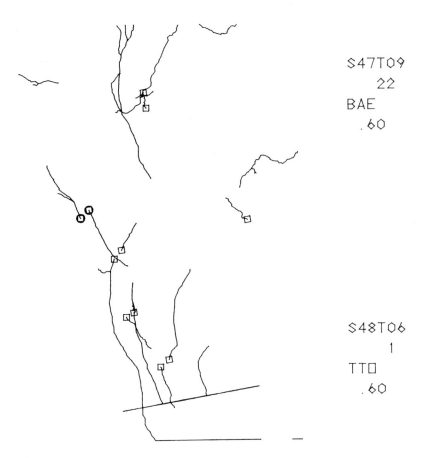

Figure 21. The CRT image during the 'merging after tracing' procedure. All the tree pieces from two adjacent tissue sections (a top section and a bottom section) are drawn on the CRT. Squares are drawn around each top tree origin and top artificial end in the bottom section and each bottom tree origin and bottom artificial end in the top section. The top section may be rotated and translated with the joystick until the squares are matched. One identified endpoint in each section is circled and information about the identified points is presented on the right side of the figure. The researcher may then push a button to signal the computer to merge the identified points.

The alternative, merging after tracing, is illustrated by *Figure 21*. All pieces of all trees are traced into the computer with no regard to their attachments. Then traced trees from adjacent tissue sections are displayed on the CRT. Squares are drawn around tree starting points and tree ending points which are at the tissue surface common to both sections. With the joystick, the researcher may rotate and translate one section with respect to the other in order to match the squares. He then presses buttons to instruct the computer to attach specific tree starting points to specific endings. The computer translates all the coordinates of each distal tree piece to attach it to its stump in the adjacent section.

7.4 Intermediate computations

When tissue is sectioned and processed further histologically, it may shrink perpendicular

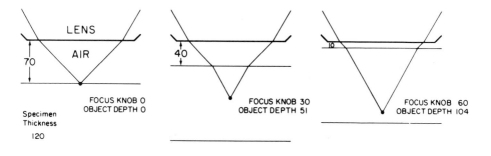

Figure 22. How unwanted refraction foreshortens focus axis readings. **Left**: a dry microscope objective lens is focused at a point on the top surface of a tissue, 70 μm below the lens surface. **Centre**: The stage is raised 30 μm by rotating the focus knob, yet because of unwanted refraction at the air−tissue boundary, an object of a 51 μm depth into the tissue is in focus. **Right**: The stage is raised 30 μm farther, to focus now on an object whose depth is 104 μm. So the focus axis knob reading is always less than reality.

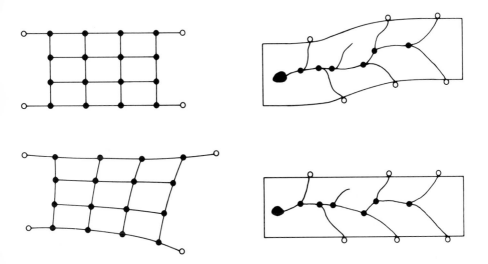

Figure 23. The tissue wrinkling problem and a solution. **Upper right**: A neuron in a wrinkled tissue section. The surface points of the neuron, marked by open circles, appear at diffent z coordinates. **Upper left**: a volleyball net whose attachment points are marked by open circles. **Lower left**: The net is deformed by moving its right two attachment points. The knots in the strings which form the net 'relax' into new positions of equilibrium. **Lower right**: The surface points of the neuron are similarly moved to their correct z coordinates, and the branchpoints on the neuron 'relax' into new positions of equilibrium.

to its plane of sectioning, the microscope's focus axis. Also, due to unwanted refraction in the microscope optical system, focus axis readings are usually less than reality. This latter problem is illustrated in *Figure 22*. These two problems sum to give focus axis readings which are perhaps 60% of reality. Therefore, to perform an accurate reconstruction, it is necessary to expand the reconstructed neuron in the z direction to return it to its thickness at the time of tissue sectioning. This may be done by a simple linear algebraic scaling.

Anatomical measurement and analysis

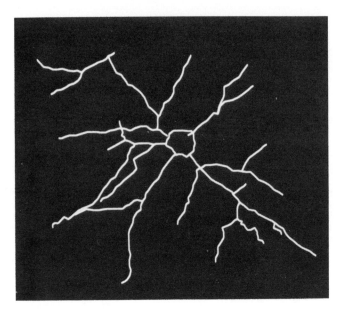

Figure 24. A stick figure diagram of the neuron drawn on a vector CRT. The figure may be rotated smoothly so that the user may better appreciate its three-dimensional structure.

Unfortunately, the tissue may also wrinkle as shown in *Figure 23*. This results in varying z readings for endpoints which are at the top of the tissue surface. It is necessary to flatten the tissue section to force all the top surface endpoints to their proper z coordinate and all the bottom surface endpoints to their proper z coordinate, while adjusting the tree so as not to distort its structure too abruptly. This dewrinkling procedure is also illustrated in *Figure 23*.

The dewrinkling algorithm handles the neuron trees like a volleyball net. Such a net is attached at four fixed corners, and the knots of the string that comprise the net relax into positions so as to balance the tensional forces upon them. If two of the corners of the net are moved, the string knots relax into new positions. Each knot's new position depends upon how far it is from the moved corners and how much the corners are moved. The gross structure of the net is distorted smoothly. The neuron is similarly distorted in z by moving each of its top and bottom surface endpoints to the exact z values of the tissue surface and the branchpoints, analogous to the net's knots, are relaxed into new positions, smoothly distorting the mathematical model of the neuron. Finally, points intermediate between the now-moved branchpoints are linearly scaled to their final positions. The result is a neuron structure scaled within a flat slab of tissue.

7.5 Three-dimensional display of the neurons

Once the neuron has been fully reconstructed in computer memory, displays may be generated as shown in *Figure 24* so that the researcher may see the three-dimensional structure of the neuron. He would like to hold the neuron in his hands and move it around in front of him as he might do if he were examining a basketball. This is, of course, not possible with a neuron. A reasonable approximation is to build a display

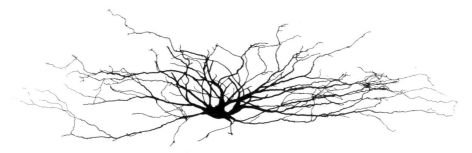

Figure 25. A realistic plot of a cat spinocervical tract neuron. The plot was made on a felt tip pen plotter in 12 frames which were then assembled and photographed. The soma is filled and the dendrites are presented in their actual thickness. Dendritic spines are varicosities and are also drawn. Modified from *J. Neurosci.* (1986), **6**, 3, cover, with permission.

```
                    Numeric summary
Text:   Golgi-stained cell from cat neocortex.  Section 7, cell number 14.

Name:   ******
No. pts summarized (incl somas)       361
No. trees summarzd (incl somas)         6

MTO TTO BTO      CP   FS   SB   BP   NE   ES  MAE  TAE  BAE  SOS  SCP SOE
  5   0   0    275    8   14   16   13    1    1    4    2    1   20   1

Limits in X,Y,Z (mu)             -105.0   102.0  -105.0   70.5  -27.0   16.0
Total fiber length (mu)          1263.9
Membrane surface area (mu**2)    8046.5
Cell volume (mu**3)              4829.6
Soma cross sectn area (mu**2)     325.4
Center of fiber length (mu)        -8.3   -13.4    -4.8
Center of surface area (mu)        -3.7   -13.5    -5.2
```

Figure 26. A numeric summary of the neuron of *Figure 19*. A soma and five dendrites have been entered into the computer. The number of each type of point and several other parameters are presented in order to characterize its morphology mathematically.

of the neuron on the CRT and allow the viewer to position it smoothly and instantaneously with the joystick. The computer terminology for such instantaneous positioning is 'real time rotation'. The combination of the user's senses of vision and kinesthesia produces for him a dramatic three-dimensional effect.

7.6 Plots of the neuron structure

A realistic plot of a neuron, as shown in *Figure 25*, may be made on a felt tip pen plotter or on a laser printer. The plots can be made from any angle, to show, for example, a sagittal view of a neuron which was traced from transverse sections. Because there are not the time constraints of real time rotation, many details can be plotted. Many neurons can be plotted on a single sheet, showing their correct relative position in the tissue. Colour can be used to highlight features or to distinguish one cell from another.

7.7 Statistical summaries of individual neurons

With reasonable effort, it is possible to generate mathematical summaries of the neuron's structure from the computer-stored model. Hundreds of summaries are possible, so

Anatomical measurement and analysis

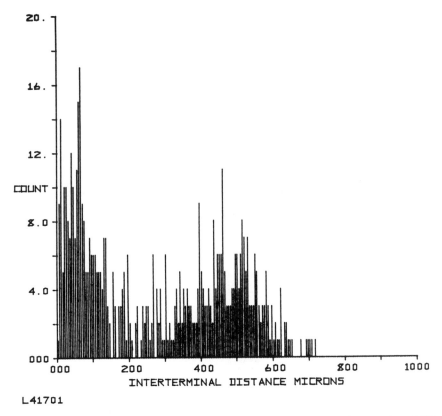

Figure 27. A histogram of the distances between the terminals of a neuron. The large number of 500 μm distances indicates that there is indeed a clustering of terminals for this cell. Reproduced from the *Brain Theory Newsletter* (1978), **3**:3/4, p. 182, with permission.

only a few usual ones which are unique to neurons will be described. *Figure 26* presents typical numeric values which are used to describe a neuron.

Figure 27 indicates the amount of clustering of the terminals of a neuron. If the terminals were spread at random as grass seed might be scattered on a lawn, no 'hump' in the middle of the graph would be expected. The existence of the hump implies a great degree of clustering.

Figure 28 presents a summary of the branching directions of a neuron. This graph is commonly used to determine where the dendrites grow.

7.8 Population summaries

As for any anatomical structure, statistical summaries may be generated in order to compare one population of neurons to another. *Figure 29* illustrates this technique. Here, for many cells, it is shown that one population has a greater membrane surface area than another and that the maximum difference in surface areas occurs near the soma.

J.J. Capowski

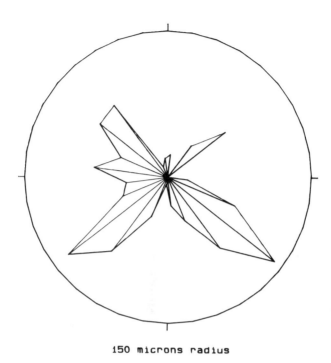

Figure 28. A polar diagram in the *x,y* plane for the neuron of *Figure 19*. Twenty four wedged-shaped bins are defined and the length of each dendrite which falls in each bin is accumulated. For each bin, a line whose length is proportional to the dendritic length in the bin is plotted from the centre of the diagram to the outside of the bin. An envelope is then drawn around the bins. This graph shows that most of the dendrites in this cell grow toward four o'clock and toward eight o'clock.

8. THE USE OF VIDEO

In the last decade, TV cameras and monitors have been coupled with computers in order to construct instruments for making anatomical measurements. The rest of this chapter will describe some of the typical ways that these are used. First however, it is necessary to describe the relevant television hardware.

8.1 Computer–television hardware

A closed-circuit television system consists of a TV camera (Dage-MTI) which observes a scene and a TV monitor which is electrically connected to the camera and which reproduces the scene. The image of the scene is focused on the front surface of the TV camera's CRT which is coated with a substance whose electrical resistivity is proportional to the amount of light incident on it. The camera sweeps an electron beam over this front surface and the monitor, in synchronization, sweeps its beam over a light generating phosphor coating. The camera generates an electrical signal which

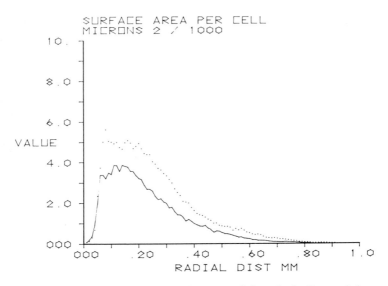

Figure 29. The membrane surface area per neuron for two populations of cells. One population, represented by dots, has a larger membrane area than the other, especially near the neuron somas. Reproduced from *J. Neurosci.* (1986), **6:3**, p. 667, with permission.

consists of two components: synchronization pulses and analogue video information. Synchronization pulses maintain the two electron beam sweeps in corresponding positions. The video signal is proportional to the light level at any instant at the point in the image where the electron beam is currently passing.

A special purpose computer board called a 'TV digitizer' or 'frame grabber' (Imaging Technology Inc.) may be plugged into the computer and may intercept the camera's electrical output. It can grab an image by sampling the video signal about 500 times per horizontal line, converting each analogue sample to a digital value, and storing each sample in a memory. This sampling is done for each line, resulting in a video image which is stored in the board's memory as an array of 500 by 500 pixels. Typically, each pixel contains 8 bits and has a numeric grey level value which ranges from 0 (for the blackest level) to 255 (for the whitest level). This digital memory is called a 'frame buffer'. The computer may read the numeric grey level of any element of the picture. It may also write a value into the frame buffer, thereby modifying its grey value. The TV digitizer board also generates synchronization pulses and video information for a TV monitor so that the monitor can display the contents of the frame buffer as it is filled from the TV camera or as it is filled or modified by the computer. This is a very powerful piece of equipment which can be used in many ways, some of which are illustrated below.

8.2 Cursor mixing using video

In any kind of tracing, it is necessary to pass a cursor over the structure to be traced and to have the computer keep track of the cursor's coordinate. In the systems described earlier, mixing the cursor with the image was done optically. It is also possible to use

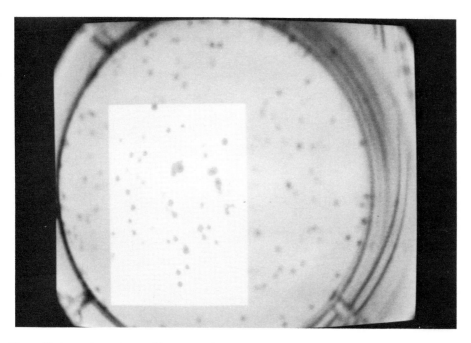

Figure 30. Improving an image with contrast enhancement. This is a television image of a group of hamster ovary cells stained with cresyl violet. Clusters of cells are apparent and should be counted. The inset shows a region of the image enhanced by expanding its grey level range. Courtesy of J.A.Cidlowski, UNC Department of Physiology.

television hardware to perform the mixing task.

The TV camera is mounted on the microscope and its video output is fed into the TV digitizer board. A data tablet, mouse or joystick is attached to the computer so that the computer may sense its position. The frame grabber continually captures the image into its memory and the computer continually reads the data tablet for its cursor coordinate. The computer loads a dot or cross into those frame buffer memory locations which correspond to the position of the data tablet cursor. A TV monitor presents the resulting microscope image modified by the computer to include the cursor (and possibly other computer-generated information). *Figure 4* shows an example of this. As the user moves the cursor over the tablet, the cross moves over the screen. The computer keeps track of the cursor coordinate and may use the coordinate to capture points in the outline as the user traces an outline.

8.3 Image enhancement

A televised image is usually inferior to the original image. This is particularly true of images which are televised, digitized into a frame buffer and then regenerated on a TV monitor. Spatial resolution and dynamic range are reduced by this image processing. However, it is sometimes possible to use a computer to shift the distribution of the information in an image so that the human visual system will glean more understanding from a certain spatial part or a certain range of grey levels of an image, at

Anatomical measurement and analysis

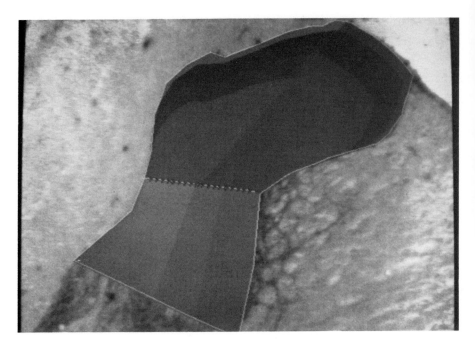

Figure 31. An example of densitometry of an image. In *Figure 4*, the dorsal horn of the spinal cord was outlined with a mouse. The program now divides the dorsal horn into 20 regions. The grey level of each pixel in a region is counted and the average grey level per pixel is computed. This value is a measure of the number of stained fibres in the region. The average grey value of each region is displayed throughout the region to provide feedback that the program has proceeded correctly.

the cost of losing information in other spatial portions or ranges of grey. If the lost information is not relevant, then there is a clear gain to the observer. The process of mathematically manipulating an image to aid the viewer in extracting more information from a part of the image is called 'image enhancement'.

Many kinds of image enhancement techniques may be applied to an image. Three will be discussed here. The first, contrast enhancement, is illustrated in *Figure 30*. The original image is digitized into a frame buffer so that the digital range of grey levels is from 0 to 255. The relevant information in the image is contained in a limited grey level range, for example from 100 to 150. So the computer program may enhance the contrast of the range 100 to 150 by shrinking the blacker range (the values of range 0 to 100 will be scaled to the range 0 to 50), shrinking the whiter range (the values of range 150 to 255 will be scaled to the range 200 to 255), and expanding the middle range (all values of range 100 to 150 will be scaled to the range 50 to 200). This will assist the viewer in picking out information whose grey levels are in the relevant middle range of the original image (Media Cybernetic Inc. Software).

A second type of image enhancement is called 'image averaging'. The information in an image may be at a very low light level, so low as to be undetectable to the observer. This occurs for example, when viewing neurons stained with fluorescent dyes which emit a very small amount of light. If a series of images of the same scene are digitized and are summed, pixel by pixel, then the random noise in the images is

J.J.Capowski

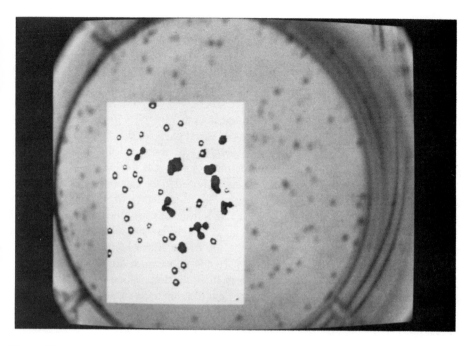

Figure 32. The counting of particles in an image. The computer program has counted the cells in the inset of the image of *Figure 30*. Those particles which fall within a specified grey level range are coloured black. Those which also fall within a specified size range are counted and so indicated with a white dot in their middle. Courtesy of E.M.Johnson, UNC Department of Physiology.

cancelled, but the non-random structural information is accumulated, so the resulting image is an enhanced, brighter view of the original, low light level image.

A final type of image enhancement, pseudo colouring, is also widely used. A high resolution black and white TV camera is much less expensive than a colour camera of the same spatial resolution. Since our eyes are very astute at discriminating colours and since colour television monitor technology is well understood and inexpensive, it is beneficial to present grey level images in a range of colours by assigning certain colours to certain values of grey. Furthermore, if the colours are selected intelligently, special effects can be achieved. For example if, in a digitized image of a CT scan, it is known that a tumour always has a certain narrow range of grey levels, the computer may specify that this range of greys be displayed in a bright red colour, to make it very easy to see.

8.4 Extracting numerical information from an image

When the image is stored in the frame buffer, the computer can, of course, extract numeric information from it. The optical density, or 'brightness' of a region may be calculated easily by adding up the grey values of all the pixels within the region. This might yield, for example, a measure of how many darkly labelled neurons exist in a region. *Figure 31* shows such an example.

Particles may be counted automatically by the computer. Here the program searches for any pixel whose grey level is within a certain range. When it finds one, it searches for any adjacent pixel which is also within the grey-level range. If the second one exists, a cluster of pixels is defined, and further adjacent pixels are sought. When all clusters are found, each is tested to see if it is within a size range. Each cluster of pixels which is neither too big nor too small is counted as a particle. As shown in *Figure 32*, feedback is provided on the TV monitor so that the user can see which particles the computer has counted. An editing feature is also provided so that the viewer can reject particles which the computer has counted incorrectly or insert particles which the computer has missed. The count of particles may be included in more advanced mathematical summaries, such as a histogram of the number of particles versus their size.

Figure 32 shows an automatic particle counter applied to the counting of cells. Another application frequently used in neuroanatomy is the counting of silver grains. In a technique called autoradiography, radioactive substances are injected into neuronal pathways, where they migrate throughout the cells. The substances are not visible, but they emit beta particles or electrons. If the top of the tissue section is flooded with a photographic emulsion, then the beta particles will expose the emulsion thereby presenting little particles of silver halide in the emulsion. When these particles are observed through a microscope using dark field illumination (the light rays illuminate the specimen from an oblique angle and only light scattered from the specimen reaches the eyepieces), they appear like bright stars in a night sky. These may be easily counted by an automatic particle counting program.

Other structures whose shape is more complex may also be recognized and counted automatically, with varying degrees of success. The counting of the various types of chromosomes is a good example. In the simplest counting program, a template or mask is defined. It is a black and white picture of a chromosome, say 20 by 20 pixels, with each pixel reduced to a single bit, either on or off. An image which contains chromosomes is digitized into the frame buffer, and each pixel is reduced to a single bit, either black or white. Then the matrix of pixels in the mask is compared to the upper left 20 by 20 matrix of pixels in the frame buffer to see if there is a reasonable match. If so, a chromosome is counted. If not, the mask is compared to the matrix of pixels one column to the right of the previous matrix. The mask is compared to every 20 by 20 matrix in the frame buffer. Because these structures are not round, it is necessary to try the mask in many different rotational orientations as well as over the entire image. So the time required to count the structures may be significant, but the counts may be more accurate than those done by a human.

10. CLOSING COMMENT

The use by anatomists of computers to extract numerical and pictorial information from structures has opened up a relatively new field, driven as much by technology and cost as by the scientist's intellectual curiosity. It has been very interesting for me over the last decade to help develop these new techniques and watch them be accepted by anatomists as the tools of their trade. The future will accelerate the change. More powerful computers at a lower cost will make it possible to extract more information from video images, allowing scientists to see and measure microscopic things which are now too difficult or too expensive to observe.

11. ACKNOWLEDGEMENTS

I would like to thank Edward R.Perl, the chairman of the UNC Department of Physiology for his financial and moral support over the years. Miklos Rethelyi, 2nd Department of Anatomy, Semmelweis University Medical School, Budapest, and a frequent visitor to our department, has spent much time and effort as critique of my efforts. William L.R.Cruce, Department of Neurobiology, Northeastern Ohio Universities College of Medicine, Rootstown, Ohio, has been most helpful and has been a cheerleader of sorts. From my department, Ellen M.Johnson and C.William Davis have been valuable backboards for technical problem solving. Most recently, I am indebted to Eutectic Electronics, Inc., Raleigh, North Carolina for providing grant support to my laboratory so that I might continue my research.

CHAPTER 4

Digital image processing and analysis techniques

GRANVILLE V.MOORE

1. INTRODUCTION
1.1 Basic principles of digital image manipulation

The manipulation of digitized images by computer is a rapidly expanding field of computer science and has many biomedical applications. This chapter describes a number of simple image manipulation techniques and basic hardware requirements for image manipulation. These are introduced in a non-mathematical manner wherever possible, and are illustrated in a description of a simple blood cell analysis system.

Computerized image manipulation may be divided into four basic areas which have different starting points and different objectives. These are image processing, image analysis, image restoration and image reconstruction.

1.1.1 Image processing

Image processing is the alteration of an image so as to produce a new image which is in some way better, for a particular application, than the original. This involves modifying existing images and is frequently used to enhance some particular aspect of an image, for a human observer. An example of this would be the use of colour, or the enhancement of edges in X-ray images, to allow clinicians to see details of an image more easily. Image processing is used in many biomedical imaging applications, and also as part of image analysis systems.

1.1.2 Image analysis

Image analysis is the process of examining an image to extract some quantitative or qualitative information from it. Most image analysis tasks require some image processing before analysis can take place. This is known as pre-processing and is, in this case, performed to enhance the aspect of the image which is to be analysed by the machine. The applications of image analysis are mainly in areas where large numbers of images need to be routinely examined, such as cell counting applications. In these situations, human operators become prone to mistakes, and appreciable variation may be found between the performance of different operators, or of the same operator working at different times. The use of computer image analysis, in these cases, assures repeatability and reliability, and may decrease processing time substantially.

1.1.3 *Image restoration*

Image restoration is the process of restoring a damaged or imperfect image to its original state, in cases where something is known about the form of degradation which has taken place. The faults which are to be corrected may be a result of defects in the imaging or digitization system, or may be due to degradation of the image due to its storage or transmission. One possible biomedical application of image restoration would be the de-blurring of X-ray images, in cases where this is caused by movement of the subject.

1.1.4 *Image reconstruction*

Image reconstruction is the process of building a two-dimensional image or set of images from data obtained in some other format, such as projections obtained from computer assisted tomography (CT or CAT scanning) or magnetic resonance imaging (MRI or NMR scanning). The use of computers in these areas is vital, so that the process may be performed in a reasonable time.

This chapter is mainly concerned with the first two of these areas, but many of the techniques described are also applicable to image restoration and reconstruction.

2. IMAGE MANIPULATION

2.1 **Image storage and representation**

In order that an image may be processed or analysed, it must be placed in a portion of the computer's memory, and this requires that each point in the image be represented as a numeric value or grey level. The process of converting an image into this form is referred to as digitization. This involves splitting the image into a large number of small regions known as picture elements (or pixels) and assigning a discrete grey level to each pixel. In most systems, the image is divided into an array of square or rectangular pixels, as shown in *Figure 1*. This array of numbers may be examined by the computer and altered or measured to produce a new image, or some numeric results. It is conventional, in image processing and analysis, to number the pixels upwards and rightwards from the bottom left-hand corner of the image, as in the axes of a mathematical graph. This is in contrast to some computer graphics systems, which number pixels downwards and rightwards from the top left-hand corner of the image.

The amount of memory storage available for each pixel determines the number of grey levels that may be used in an image. If n bits of storage are allocated to each pixel, then each may take values from 0 to $2^n - 1$. It is conventional to use the value 0 to represent the lowest intensity, black, and $2^n - 1$ to represent the highest intensity, white. Typically, images may be represented using 6 or 8 bits per pixel, giving 64 or 256 grey levels respectively.

If an image is stored using only 1 bit per pixel, only values of 0 or 1 may be used to represent the intensity level at each point. These, therefore, represent full black and full white respectively, with no intermediate grey values. Such an image is referred to as a binary image, and these are frequently used in image analysis to mark or represent regions of interest. Images which do contain intermediate grey values are known as grey-scale images, to distinguish them from binary images. The number of available

2	2	2	4	4	4	5	6	7	7	8	9	10	11	12	13
3	3	5	4	6	5	6	6	7	8	8	9	10	11	12	12
4	5	8	7	7	4	6	6	7	8	9	10	10	11	12	12
9	7	9	8	8	8	7	7	8	9	9	10	11	11	11	12
7	15	12	11	15	21	18	10	8	10	11	10	11	11	11	11
13	20	22	15	22	25	20	21	20	11	12	11	10	11	11	11
24	25	26	24	33	31	15	22	21	15	14	12	12	10	10	11
34	32	29	33	44	43	32	24	20	32	21	20	14	12	11	10
36	31	41	44	46	45	45	44	34	28	31	26	21	15	12	11
38	42	44	46	48	47	48	45	45	39	32	20	21	18	13	14
40	44	47	48	49	49	47	48	46	36	31	31	23	21	17	17
41	45	49	49	50	50	52	50	47	46	34	27	31	24	18	19
43	45	51	50	50	50	51	49	48	47	35	31	28	23	24	23
46	45	47	50	53	51	50	49	48	47	45	36	36	28	21	25
38	40	46	48	50	50	49	50	49	48	45	37	32	27	24	25
35	41	44	46	49	49	49	47	45	46	39	40	37	35	33	28

Figure 1. Image representation as an array of pixels.

grey levels in a grey-scale image is known as the grey-scale resolution of the image.

A second important characteristic of stored images is the spatial resolution. This is the number of points of the original image which are sampled to make up the stored image, and is, thus, the number of pixels in each dimension of the image. Currently available systems use images which range from 256 × 256 pixels to 4096 × 4096 pixels, and, typically, an image processing system might use a 512 × 512 image with 64 grey levels (256 K 6-bit pixels). From this it may be seen that images generally require a large amount of storage space, and so it may be necessary to add extra memory capacity to some computers to deal with images. Such hardware requirements are discussed later in this chapter.

2.2 Colour images

Monochrome images may be represented by a simple array of integers, but, for colour images, it is necessary to use additional memory to store colour information. This is usually done by splitting the image into red, green and blue components, and storing each component as an individual image. Red, green and blue are known as primary colours, since, by combining these colours in the appropriate proportions, a mixture may be obtained which is, to the human eye, indistinguishable from any specified colour.

It should be noted, however, that this is merely a feature of the human visual system, and it is quite possible to construct imaging systems which are directly sensitive to other wavelengths of electromagnetic radiation, including visible light, infra-red, ultra-violet or even X-rays.

Digital image processing and analysis techniques

It should be noted that it is not possible to store colour images using the technique of storing a small number of 'colour numbers' and mapping these in a separate memory unit to give a relatively small 'palette' of colours selected from a large range. Such devices are widely used for computer graphics, but these rely on the property that only a small number of the many possible colours will be in use at any time. This is true of computer-generated graphics, but not of 'natural' images. Although colour information is useful in many applications, this chapter is concerned mainly with the processing of monochrome, or grey-scale images.

2.3 Neighbourhoods

The pixels surrounding any point of interest are known as its neighbours, and directly adjacent pixels are known as immediate neighbours. Frequently, these neighbouring pixel values are used, in conjunction with the value of a pixel, to calculate a new processed value for that pixel. In the rectangular tesselation shown in *Figure 2*, the immediate neighbours of the pixel, x, fall into two categories, those connected by sides (marked a), and those connected by corners (marked b). It should be noted that the centres of the pixels marked a are, in fact, closer to the centre of x than those marked b. For this reason, some processing operators give preferential treatment, or weighting, towards the side-connected neighbours.

2.4 Connectivity

For some aspects of image analysis, it is sometimes required to determine whether two pixels are connected to each other by an unbroken line of pixels of some value. In order

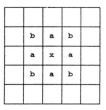

Figure 2. The neighbourhood of a pixel.

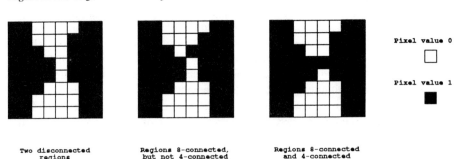

Figure 3. Connectivity of regions.

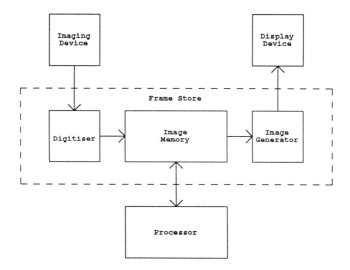

Figure 4. Basic image processing hardware requirements.

to decide this, it is necessary to define what constitutes a connection between two pixels. For some purposes, all eight neighbours are considered to be connected to a pixel, and this is known as an eight-connected scheme. In other cases, it may be that only the four side-connected neighbours are considered to be connected, and this is referred to as a 4-connected scheme. These are illustrated in *Figure 3* where it should be noted that the pixels of value 1, which denote the regions of interest, are shown for the purpose of printing, as dark pixels. This is the reverse of the expected appearance on a video screen, where such regions would usually be displayed in white.

3. BASIC HARDWARE REQUIREMENTS

We may break down our hardware requirements into six items, as shown in *Figure 4*.

(i)   An imaging device to convert the optical image into an analogue electrical form.
(ii) A digitizing device to convert this analogue image into a digital form.
(iii) A memory unit to store the image during processing.
(iv) A processor to manipulate the image.
(v)   A display generator to extract the processed image from memory.
(vi) A display device to view the processed image.

In most systems, the digitizer is combined with either the imaging device or the image memory device, to form a single unit. Similarly, the display generator is usually incorporated into the image memory unit, and this combination is referred to as a frame store. It should, however, be noted that, although most commercially available frame stores do contain a digitizing unit, this is not always the case.

3.1 Image input devices

To allow an image to be stored and processed in the manner described above, the optical

Digital image processing and analysis techniques

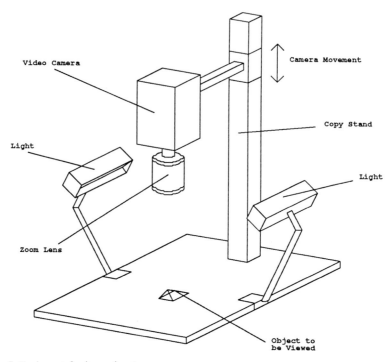

Figure 5. Equipment for image input.

image must first be converted into an electrical form, which may be suitably digitized. This may be done using any one of a variety of devices, but the video camera is, by far, the most commonly used device for general-purpose image processing and analysis.

3.2 Video cameras

The video camera is popular because it is flexible, easy to use and is relatively cheap. It is suitable for the digitization of existing two-dimensional images, such as photographs, for digitization of images of three-dimensional objects or scenes, and it may also be attached to various pieces of laboratory equipment, notably the optical microscope. Video cameras do not usually produce a direct digital output, and so an additional digitizing device must be attached to them. As mentioned earlier, this is often included in a frame store unit.

Video cameras may be subdivided into two types: those which contain a vidicon or similar vacuum tube (similar in many respects to a cathode ray, or television, tube), and those which use charge coupled devices (CCDs). Traditionally, vidicon cameras have been cheaper and have been capable of higher spatial resolution than CCD cameras (typically $512 \times 512 \times 6$ bits), but recent advances in CCD technology mean that comparable resolution may soon be attainable with these devices. The advantages of CCD cameras are that they are compact, relatively rugged, and are capable of good grey-level resolution (up to 8 bits).

In addition to the camera itself, which should be of a type suitable for use with interchangeable lenses, a certain amount of other optical equipment is required for image input (*Figure 5*).

3.2.1 *Lenses*

Ideally, a selection of lenses should be available, but, for most purposes, a good-quality 25–80 mm zoom lens with aperture range up to f4 is sufficient. It is usually necessary to use some form of adapter to allow the required lens to be fitted to the camera.

3.2.2 *Extension tubes*

These are relatively inexpensive hollow tubes which fit between the lens and the camera to allow close-up focusing, where the object to be imaged is less than about 20 cm from the camera. These are useful for examining small objects, or for digitizing photographic slides.

3.2.3 *Camera stand*

Some form of mechanical support is required for the camera, and, for this, a photographic copying stand is frequently used. This device allows flat photographs or small objects to be placed on the base plate and the camera moved vertically to obtain the desired view. More expensive stands allow lateral camera movement, and may allow the camera to be swivelled to a horizontal position to digitize three-dimensional scenes.

3.2.4 *Lighting equipment*

Lighting is a crucial part of the imaging process, and may require much experimentation to obtain satisfactory results. In general, diffuse light is preferable to point-source illumination, to avoid undesirable reflections from the object to be digitized. This problem is particularly noticeable when imaging metal objects, or gloss-finish photographs. A light-box, as used by photographers for viewing photographic slides or negatives, is a useful tool for image input. This may be placed on the base of the copy stand so that small objects may be silhouetted against a bright background. A light-box is also required if transparencies are to be digitized.

3.3 **Other image input devices**

A number of other input devices exist, but many of these are specialized towards specific tasks. The most commonly used general-purpose device, other than the video camera is the microdensitometer. This is a mechanical device in which a photograph or similar flat image is scanned by moving it beneath a single fixed sensor. The scanning may take place using motors to move the image in the horizontal and vertical directions or, more commonly, by fixing the photograph to a rotating drum and scanning slowly along the axis of the drum. Digitization may take place to very high spatial and grey-level resolution ($>1024 \times 1024 \times 8$ bits). Such devices do, however, have several disadvantages; they can only be used to deal with flat images, setting up the equipment is time-consuming, and the scanning process is, itself, slow. Microdensitometers may produce either analogue or digital outputs, and so may require a separate digitization

Digital image processing and analysis techniques

unit, but as the image scanning is slow, this may be a relatively inexpensive analogue-to-digital converter.

3.4 Image display devices

The normal method of displaying images is to use a standard high-resolution interlaced video monitor. Cathode ray tubes with long-persistence phosphors may be used in these monitors to reduce the 'flickering' effects caused by the scanning of the electron beam used to generate the displayed image. On a good quality monochrome monitor, images of up to 1024 × 600 × 8 bits may be displayed. The vertical resolution of these raster-scan devices is limited by the number of scan lines in the raster, and, although monitors are available which use 1024 horizontal lines, these are very expensive. Screen displays based on liquid crystal technology may offer a useful alternative in the future.

Hard copy of images may be obtained in a selection of ways. The simplest of these is to use a conventional photographic camera pointed at a monitor in a completely darkened room. Long exposure times (0.5 – 1 sec) are necessary to avoid unevenness in brightness caused by the output scanning. One major problem with this method is the 'barrelling' effect caused by the curvature of the monitor screen, but this may be reduced by using a long focal length lens at a large distance from the screen. Special flat-screen monitors are available for this purpose, but these are expensive, and, in any case, great care must be taken to align the optical axes of the monitor and the camera.

Other hard-copy image output devices include some similar to the microdensitometer, which may be used to expose photographic film using a single-point light source, scanned across the film. Electrolytic paper, which turns dark when an electric current is passed through it, may also be used in similar equipment. Most dot matrix printers offer at least binary image output, and many are capable of generating up to 16 grey levels by using varying dot sizes, and by overprinting. Colour printers are also available, but the quality obtainable is substantially inferior to that produced by direct photography.

3.5 Frame stores

Many computers used for image processing and image analysis do not have sufficient built-in memory to store an image (typically 256 Kbytes). In such cases, an external frame store, as mentioned earlier, is used to store the image, and this is accessed by the computer through an input/output port. On larger machines, processing may take place in the main store of the computer, using the frame store only for digitization and display. A basic frame store is composed of a section of memory with suitable hardware attached to load this with an image (usually in a single frame-time) from an input device, and to repeatedly read out the stored image to refresh the display screen.

The frame memory is usually organized as a set of bit-planes, each $2^n \times 2^n \times 1$ bit. In this scheme, one bit from each plane is taken to construct each pixel value (*Figure 6*). To store a binary image, only a single bit-plane is required, but most frame stores have at least four, and more typically eight, bit-planes. This organization allows a frame store to be used to hold a number of images of different grey-scale resolution simultaneously. For example, a frame store with eight bit-planes could store a 6-bit image and two binary images simultaneously. Frequently, it is possible to arrange that

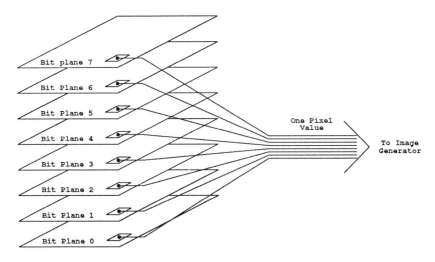

Figure 6. Bit-plane organization of a frame store.

the displayed image is a combination of all the stored images, so that regions may be highlighted, or text may be displayed, by overlaying binary images over a grey-level image.

3.6 Processors

Processors used for image processing and analysis vary in size and power, from personal computers to supercomputers such as the Cray 2 or Cyber 205. The time taken to process an image depends on the complexity of the operation to be performed and the available hardware, and so this may range from hours down to less than one second. To relieve this problem, where very large machines are not available, some systems use additional hardware designed specifically for image processing. Many powerful specialized machines have been constructed for image processing and image analysis research, but most commercially available add-on units tend, however, to be either expensive or somewhat limited in their applications.

4. SIMPLE IMAGE PROCESSING TECHNIQUES

As has been stated earlier, image processing operations are designed to display images in such a way that the best possible use can be made of the existing information in the image, and a number of simple techniques to do this are now described.

4.1 The grey-level histogram

The grey-level histogram is an important tool in image processing. This is a conventional histogram, with grey level plotted on the abscissa and number of pixels on the ordinate. This is usually represented in the computer as an array of numbers in which the Nth element contains the number of pixels in the image whose grey-level value is N. Thus, dark images produce grey-level histograms with large values at low N,

Digital image processing and analysis techniques

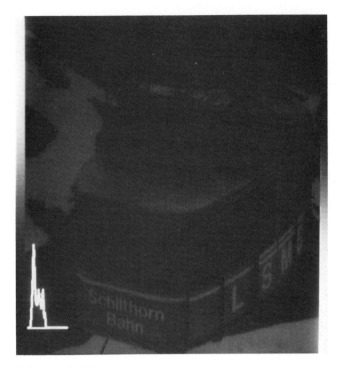

Figure 7. A dark image, with histogram.

Figure 8. A light image, with histogram.

Figure 9. A low-contrast image, with histogram.

as shown in *Figure 7*, and light images produce large values at high *N* (*Figure 8*).
The width of the histogram is closely related to the contrast of the image. Low-contrast images, such as in *Figure 9*, have histograms which are sharp and tall, whereas well-contrasted images, such as *Figure 10*, have histograms which are wide, tapering to zero only at the extreme values of grey level. Clipping is an effect which is caused by the presence of too much contrast in the image, so that the information at the extremes of the intensity range are lost, since these values lie outside the digitized range. This results in large, sharp peaks at the absolute extremes of light or dark, as seen in *Figure 11*.

4.2 Point transformations

Both clipping and lack of contrast result in images which are difficult to interpret. In the former case, there is no way to restore the lost information, other than by re-digitizing the image. In the latter case, however, we may expand the contrast by applying a point transformation to the image. This does not increase the amount of information in the image, but does make the existing information more visible.

In a point transformation, all pixels with any particular grey level are changed to the same new level, according to a transformation function. This function is normally expressed as a table of corresponding old and new pixel values, and each pixel in the image is replaced by the value from this table which corresponds to its old value (*Figure 13*). Given the histogram of an image and the transformation function, it is possible to predict the histogram which will result from the new image, and so we may design

Digital image processing and analysis techniques

Figure 10. A correctly digitized image, with histogram.

Figure 11. An over-contrasted image showing clipping, with histogram.

Figure 12. The low-contrast image of *Figure 9*, after contrast expansion.

transformation functions which will produce an image with a flatter histogram. This image should then be easier to view, since it will have correct contrast.

To expand the contrast of an image such as *Figure 9*, we may simply cause the non-zero section of the histogram to be stretched across the full range of grey values, leaving large spaces between the non-zero elements. This is known as histogram expansion, and the function, illustrated in *Figure 13*, has this effect. The resulting contrast-expanded image is shown in *Figure 12*. It should be noted that this operation never removes any information from the image, and the original may be exactly restored by an inverse transformation. If, however, the histogram of the initial image has a large non-zero section, this form of histogram manipulation may not result in a large degree of perceived improvement.

Ideally, we require a transformation such that the output histogram will be as close to linear as possible. This would mean that the the pixel values in the image were distributed equally amongst the available grey levels, and, using information theory, it is possible to show that this maximizes the information retained in the image. In practice, since the histogram is a discrete function, we can only approximate to a linear histogram. The process of altering an image to achieve this objective is referred to as histogram equalization, and is discussed in detail by Gonzalez and Wintz (1).

5. LOCAL OPERATORS

Frequently, the use of point transformations is insufficient to enhance the desired features in an image, and it is necessary to use a more powerful method. A local operator is

Digital image processing and analysis techniques

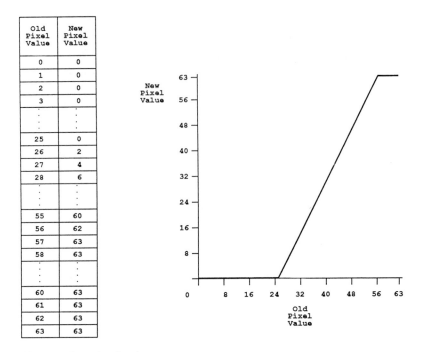

Figure 13. A transformation function.

a technique in which the new value of a pixel is based not only on its old value, but also on those of its neighbours (*Figure 14*). This may be just the immediate four or eight neighbours, or may include a larger 'window' of the image. Obviously, the larger the neighbourhood which is used, the more information is available. However, the examination of large neighbourhoods may be time-consuming, and so the most commonly used local operators use 3 × 3 or 5 × 5 neighbourhoods.

5.1 Image smoothing

Image smoothing is an operation whose effect is to 'blur' the image slightly. This may be done to attempt to reduce the effects of unwanted noise or 'graininess' in the original image, and, thus, to allow better subsequent segmentation of the image (discussed later). An example of a simple local operator for image smoothing is the 3 × 3 mean average operator. In this case, each pixel is replaced by the mean average of the nine pixels in the 3 × 3 neighbourhood centred on it. This is a very simple operator which is easy to implement and is fast to run, but, however, it is not really suitable for general use since it may tend to introduce unwanted artefacts into the image. A better operator for this type of application would be a 3 × 3 Gaussian-weighted filter, which involves taking a weighted average of the pixels in the neighbourhood, such that the pixels nearest to the centre of the neighbourhood contribute more to the final value than pixels at the edge of the neighbourhood. A commonly used weighting scheme is:

$$\begin{matrix} 1 & 2 & 1 \\ 2 & 4 & 2 \\ 1 & 2 & 1 \end{matrix}$$

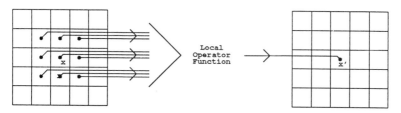

Figure 14. A local operator.

$$X = (B + 2C + D) - (H + 2G + F)$$
$$Y = (D + 2E + F) - (B + 2A + H)$$
$$P' = \sqrt{X^2 + Y^2}$$

Figure 15. The Sobel operator.

To calculate the result value for some pixel, it is necessary to multiply each of the neighbours by the corresponding weight, and to add these results together. This is then divided by the sum of the weights, to produce a new pixel value in the same range of grey levels as the original image. It should be noted that, to speed up the division process, the sum of the weights used in this case has been made to be 16, which is power of two. This means that, in a machine-code implementation, a rapid arithmetic shift may be used instead of a lengthy division.

This local operator is an example of an process known as convolution, and the matrix of weights is sometimes known as a mask. Different masks may be used to achieve different effects, but the multiply, add and divide sequence is common to all such operators.

5.2 Edge detection

A second example of a local operator is the edge detection operator known as the Sobel operator. This operator is used to detect sharp edges in an image, by replacing each pixel in the image by a value which depends on the image gradient, or the rate of change of grey level, at that point. In flat regions of the image, where pixels are of approximately uniform colour, the rate of change of pixel value is low, and so these areas become dark in the processed image. At any edges in the image, however, the rate of change of grey level is high, and so these become bright lines in the resulting image. The Sobel value for any pixel is obtained by calculating the convolution of the immediate neighbours with the two masks

$$\begin{matrix} 1 & 2 & 1 \\ 0 & 0 & 0 \\ -1 & -2 & -1 \end{matrix} \quad \text{and} \quad \begin{matrix} -1 & 0 & 1 \\ -2 & 0 & 2 \\ -1 & 0 & 1 \end{matrix}$$

These two partial results are each squared and the sum of the two is square-rooted, and used as the new pixel value (*Figure 15*). This operator has the effect of generating a smoothed gradient function, which avoids the undesirable enhancement of noise effects

Digital image processing and analysis techniques

Figure 16. Image processed by the Sobel operator.

which are associated with many other edge detection techniques. An image processed by the Sobel operator is shown in *Figure 16*.

5.3 Implementation of local operators

It should be noted that, in general, the new value for a pixel is based on the old value of the neighbourhood, and, from this, it may be seen that the image cannot be simply updated in a pixel-by-pixel manner, as this would discard information which would be required to calculate subsequent pixels. One simple method of avoiding this would be to have sufficient memory to store two images, and to build up the new processed image in the second frame store, without altering the original source image. This is, however, an expensive alternative, and a technique exists to avoid this.

To exploit this technique, we observe that, for the pixels around the edge of the image, insufficient information exists to correctly calculate a result in the same way as for the rest of the image, since these pixels do not have a full complement of neighbours. This means that an image processed by a $w \times w$ operator must necessarily be $w-1$ pixels smaller than the source image, in each direction. It may be seen that, in *Figure 17*, the corner pixel marked a0 in the source image is used in the calculation of the value of the pixel b1′, and is not used again. Once this calculation is performed, therefore, the pixel a0 may be overwritten by the new value, b1′. Similarly, the new values, b2′, b3′,..., c1′, c2′, etc. may be written into the space previously occupied by the pixels a1′, a2′,..., b0′, b1′, etc., forming an intermediate image. After the complete

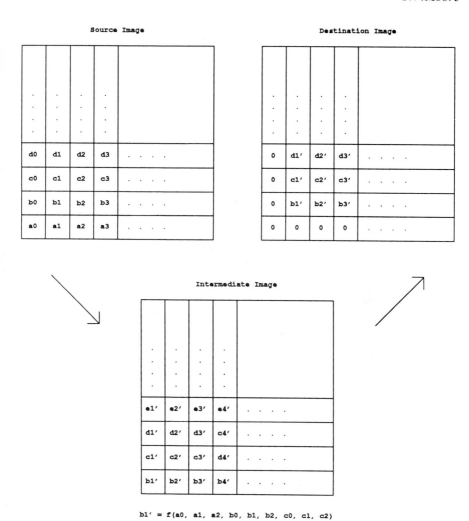

Figure 17. In-place use of local operators.

image has been processed, it is necessary to shift the intermediate image upwards and to the right, to position the new destination image in correct alignment or registration with the original.

6. IMAGE ANALYSIS TECHNIQUES

The image processing techniques which have been described so far are relatively simple to implement, since they involve performing the same, simple operation on each pixel of an image. This is, however, in contrast with image analysis operations, which are generally concerned with the measurement or identification of individual features in the image. This is not easily achieved by a simple examination of each pixel of the image in turn, and so it is frequently necessary to construct some sort of shape represen-

Digital image processing and analysis techniques

Figure 18. Freeman chain codes.

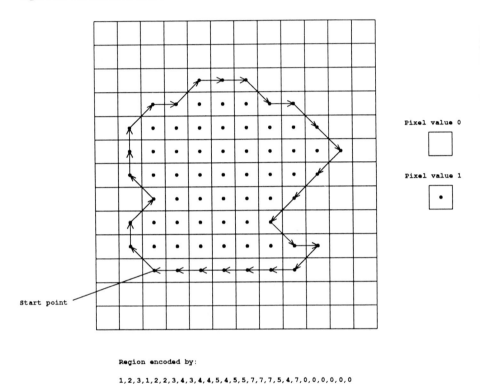

Region encoded by:

1,2,3,1,2,2,3,4,3,4,4,5,4,5,5,7,7,7,5,4,7,0,0,0,0,0,0

Figure 19. An example of the use of Freeman chain code.

tation for the features encountered. This shape representation may then be further examined for measurement or classification purposes.

6.1 Freeman chain code

One commonly used form of shape representation is Freeman chain code, often known simply as chain code. This allows an arbitrarily shaped region to be represented as a sequence of short vectors, where each vector represents a move from one pixel to an adjacent one, as shown in *Figure 18*. Chain code may be used to represent arbitrarily curved lines in an image, or alternatively, if the vectors form a closed loop, a solid filled shape or region of interest may be represented (*Figure 19*).

One algorithm to construct chain code from a binary image, where a feature composed of 1s occurs on a background of 0s, uses the principle of tracking the boundary in an clockwise sense, turning outwards towards the left wherever this is possible. This may be expressed as follows:

(i) Find a starting point, by scanning up the image, one line at a time, commencing at the bottom left corner, until a pixel of value 1 is encountered. Label this position as the 'current pixel'.

(ii) Starting at direction vector 0 (*Figure 18*), check each of the neighbouring pixels until a 1 is found. The direction of this pixel is the first vector. If no such neighbour is found, the region is simply a single isolated pixel, and is represented by an empty list.

(iii) Set a 'current direction' pointer to be this vector, and move the 'current pixel' to the end of it.

(iv) Starting with the pixel to the left (two vector positions anticlockwise) of the current direction, again check the neighbours for the first 1. This is the next vector.

(v) Repeat from step (iii) until the current point position is the same as the starting point. The chain code is then complete.

Chain code could, in a similar manner, be constructed in an anticlockwise manner, and the choice is arbitrary. It is, however, important that any system should be coherent, and use only one or the other.

6.2 Feature measurements

A number of measurements may be made from chain code representations of image features. The perimeter of a region is simply the number of vectors in the chain code list, and the area of the region may be calculated as a line integral. It is not, however, necessary to understand the detailed mathematical principles of the line integral in order to use this algorithm, which may be implemented as follows.

(i) Create variables, A, Y, N and V, and set A to zero, Y to zero, N to zero, and set V to be the first vector taken from the chain code list.
(ii) Add 1 to N.
(iii) If V is 3, 4 or 5, add Y to A.
(iv) If V is 1, 2 or 3, add 1 to Y.
(v) If V is 5, 6 or 7, subtract 1 from Y.
(vi) If V is 1 or 5, subtract 0.5 from A.
(vii) If V is 3 or 7, add 0.5 to A.
(viii) If V is 0, 1 or 7, subtract Y from A.
(ix) Set V to be the next vector taken from the chain code and repeat from step (ii).

If there are no more vectors, then the area is given by $1 + A + N/2$.

Steps (vi) and (vii) may be omitted for additional speed if lack of accuracy for small areas is acceptable. Similarly, step (ii) may be omitted, and the area approximated to $1 + A$, or simply A.

The area of a region may be required as a final result to be presented to the human operator, or it may be used as part of a classification process, to distinguish different classes of object by their size. Another characteristic which is frequently useful in such cases is the circularity of the region. This may be defined as

$$\text{Circularity} = \text{Perimeter}^2/\text{Area}.$$

Digital image processing and analysis techniques

Shape		Circularity
 ◯	Circle	12.6 (4xPi)
 ⬡	Hexagon	13.9
 ◻	Square	16
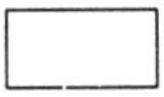 ▭	2x1 Rectangle	18
 ▭	4x1 Rectangle	25

Figure 20. Circularity values for simple shapes.

The circularity is low for regions which resemble a circle, but higher for regions which are relatively long and thin, or have 'crinkled' edges, thin peninsulas, or deep cuts. This characteristic has the advantage that it is totally independent of the size of the feature, and is purely a measure of shape. *Figure 20* shows a number of simple shapes and their circularity values.

7. A SIMPLE CELL ANALYSIS SCHEME

Image analysis, as has been mentioned, is the process of examining an image to perform some sort of measurement, or to make some sort of judgement. A number of simple techniques have been explained, and, to illustrate these, a simple example of an image analysis sequence is now described. The example selected is a commonly encountered problem; that of red blood cell counting. A digitized red blood cell image is shown in *Figure 21*. It may be seen that this somewhat noisy image contains a number of red blood cells, which appear as dark regions, and also a number of objects which are not cells. The sequence of operations used to process this example comprises four phases.

(i) Pre-processing.
(ii) Segmentation.
(iii) Feature selection.
(iv) Feature measurement.

Pre-processing is performed to remove unwanted noise from, and to generally 'clean up', the grey-level image. Segmentation is the process of dividing the image into regions, each of which represents a feature in the image which could be one red blood cell. The regions which do not represent suitable cells are discarded in the feature selection phase, and the remaining cells are measured in the final phase.

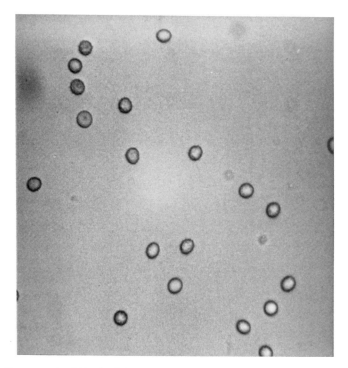

Figure 21. Unprocessed red blood cell image.

7.1 Pre-processing

The pre-processing used for this example is a Gaussian filtering operation, similar to the one described earlier. The result of this operation may be seen in *Figure 22*. The resulting image is less grainy than the original, and so the boundaries of the cells are smoother. The effects of this pre-processing may be seen in the segmentation phase.

7.2 Segmentation

Many methods exist to segment images, and many of these are suitable to separate the red blood cell image into a set of regions. The simplest of these techniques is manual thresholding. In this scheme, the grey-level image is displayed to the operator, who supplies a threshold value, and this is used to generate a binary image from the grey-level image, by using white (1) where the grey-level image is below the threshold, and black (0) where the grey-level image is above, or equal to the threshold. The operator decides whether this segmentation is suitable, and either accepts it or repeats the process using a new threshold value. *Figure 23* shows the segmented image, and it should be noted that the boundaries of the cell regions are relatively smooth. This is due to the pre-processing operator, without which the cells would have undesirable jagged edges.

7.3 Feature selection

It may be seen that many regions in *Figure 23* do not represent red blood cells, and

Digital image processing and analysis techniques

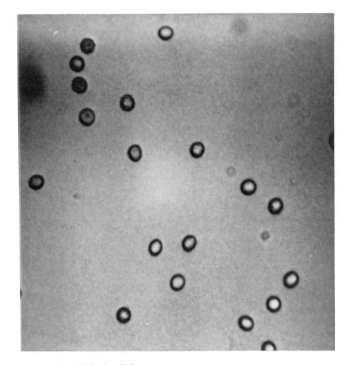

Figure 22. Pre-processed red blood cell image.

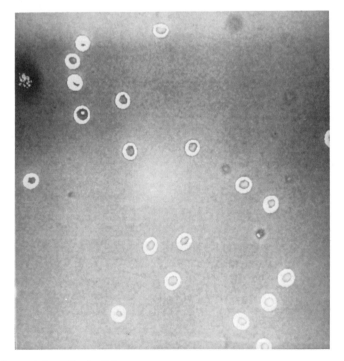

Figure 23. Segmented red blood cell image.

some thresholded cells are not suitable for counting, since they lie on the edge of the image. The regions may be classified into three groups.
(i) Red blood cells, suitable for counting.
(ii) Incomplete cells at the edge of the frame.
(iii) Small specks, caused by dirt or camera noise.

The two unwanted categories of region must be removed from the image, in order that the red blood cells may be counted. The removal of small specks may be performed in many ways, but two possible methods are:
(i) the regions could each be measured, and those less than a certain size discarded, or
(ii) a local operator could be employed to remove them from the binary image.

In this example, a local operator which belongs to a group known as morphological transforms (2) is used. This operator is known as an opening operator, and is composed of two parts, erosion and dilation. In erosion, the outer layer of pixels of each region is removed (*Figure 24*). This means that small objects which are of less than two pixels in width will disappear. The second operation, dilation, consists of growing a layer of pixels onto the outside of each object. The combination of these two operations results in an image in which large objects are relatively unchanged from the source image, but small objects are removed entirely. More powerful opening operators may be constructed by applying n erosions, followed by n dilations. The erosion operations may be implemented as a 3×3 local operator by noting that a pixel is to be set to 1 in the destination image if, and only if, it and all of its neighbours are set to 1 in the source image. This may be easily performed by using the logical AND of the entire neighbourhood to produce a new pixel value. In the case of dilation, a pixel is set to 1 in the destination image if it, or any neighbouring pixel, is set to 1 in the source image. In this case, therefore, a logical OR operation may be used. *Figure 25* shows the results of applying an opening operator, composed of one erosion and one dilation, to the segmented red blood cell image. An operation known as closing, composed of dilation followed by erosion, has the opposite effect to opening, that of filling small holes in regions.

Once small specks have been removed from the image, the partial cell must be removed from the edge of the frame. It should be noted that it is necessary to take into account the fact that such partial cells have been removed, when, eventually, the cell density is calculated and displayed to the operator. These incomplete cells may be removed by tracing around the outside border of the image, following the contours of the adjoining cells and constructing a chain code representation, as described above. To remove the unwanted partial cell regions, all pixels outside this area are cleared, leaving only the red blood cells to be counted (*Figure 26*).

7.4 Feature measurement

The operations of measuring and counting regions are central to image analysis systems, and these are often provided as built-in pieces of software with image analysis systems. It is possible to count the number of regions in an image in several ways. One possible method is to apply a series of local operators, similar to erosion, to reduce each region

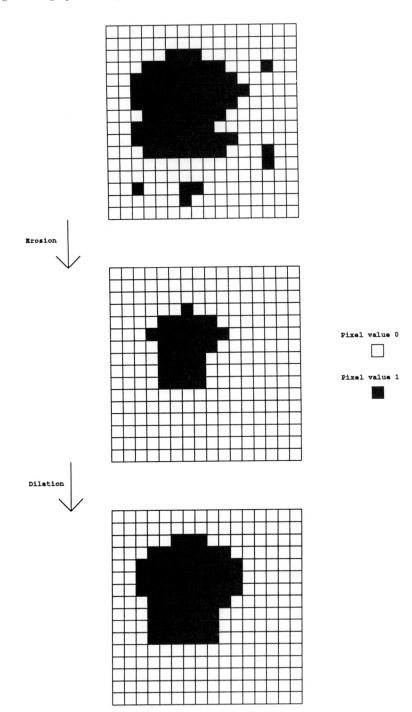

Figure 24. Erosion and dilation.

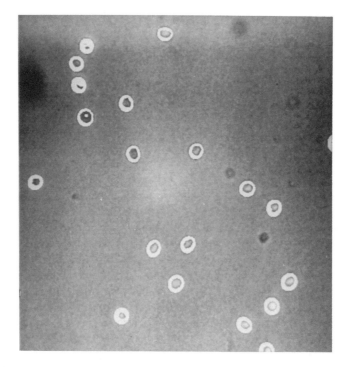

Figure 25. Segmented red blood cell image, after an opening operation.

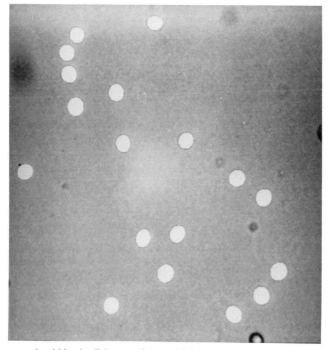

Figure 26. Segmented red blood cell image, after partial cell removal.

to a single point, and to count the remaining points. This is relatively easy to implement, but may be very slow to execute.

A somewhat faster method of region counting, which also allows some shape measurements to be derived easily, is to scan the image, one line at a time, starting at the bottom left corner until a set pixel is encountered. This pixel must be the bottommost pixel of a region, and the boundary of this region may be traced and represented using chain code. Since all that is required here is a count of the regions present, each region may be deleted after its boundary has been traced. This allows the next region to be detected in exactly the same manner as the first. The deletion is performed by resetting all pixels inside the chain-coded boundary, and this is best done by converting the region to a scan-line-based form, although other simpler, slower algorithms exist. Since, however, this process is not straightforward, it is not described in detail here.

By repeating the process of detection, tracing and removal, each of the regions in the image is, in turn, examined. It would be quite possible, at this stage, to take other measurements, and to use these as the basis for some form of decision-making process. As an example, in this case, examination of the circularity value for each region could be of use in detecting abnormal cells, since normal cells are almost exactly circular. This could be used in such a way that any region with an appropriate area for a red blood cell, but with an excessive circularity value, could be brought to the attention of the human operator, who would be able to take appropriate action.

After counting the cells, the only remaining action is to present the results to the operator. In a practical cell counting system, results would probably be stored and examined by the technician in batches, rather than for each individual frame. This does, however, require the use of a computer-controlled motorized microscope, so that the machine may move to the next frame to be processed, without the need for manual intervention. For applications such as this example, where the rate of processing is of high importance, it is highly desirable that the system should be fully automatic, and the expense of such equipment is usually justified.

8. SUMMARY

The basic principles of digital image processing and analysis have been presented, and a number of simple techniques described. These are, as is natural for an introductory text, only a very small selection from the large number of available techniques, but, with only these simple operators, practical, useful systems may be constructed, as has been shown in the example of a cell counting system.

9. FURTHER READING

More details of the algorithms and methods presented in this chapter may be found in other texts. Grey-level image processing techniques are described by Gonzalez and Wintz (1), Pratt (3) and Castleman (4), and, in more mathematical detail, by Rosenfeld and Kak (5), and Rosenfeld (6). Binary image processing and other morphological transforms are discussed in detail by Serra (2). Pavlidis (7) describes a number of algorithms for image processing and analysis, and these, together with wider aspects of computer vision and image understanding are discussed by Ballard and Brown (8) and Marr (9). Preston and Duff (10) discuss image processing and analysis techniques and describe a number of specialized machine architectures to increase processing speed.

10. REFERENCES

1. Gonzalez,R.C. and Wintz,P. (1977) *Digital Image Processing*. Addison-Wesley, Reading, MA, USA.
2. Serra,J. (1982) *Image Analysis and Mathematical Morphology*. Academic Press, New York, USA.
3. Pratt,W.K. (1978) *Digital Image Processing*. John Wiley and Sons, New York, USA.
4. Castleman,K.R. (1979) *Digital Image Processing*. Prentice Hall, Englewood Cliffs, NJ, USA.
5. Rosenfeld,A. and Kak,A.C. (1976) *Digital Picture Processing*. Academic Press, London, UK.
6. Rosenfeld,A. (1979) *Picture Languages*. Academic Press, New York, USA
7. Pavlidis,T. (1982) *Algorithms for Graphics and Image Processing*. Computer Science Press, Rockville, MD, USA.
8. Ballard,D.H. and Brown,C.M. (1982) *Computer Vision*. Prentice Hall, Englewood Cliffs, NJ, USA.
9. Marr,D. (1982) *Vision*. W.H. Freeman and Company, San Francisco, CA, USA.
10. Preston,K. Jr. and Duff,M.J.B. (1984) *Modern Cellular Automata*. Plenum Press, New York, USA.

CHAPTER 5

Computers in cardiac research

E.SKORDALAKIS and P.TRAHANIAS

1. INTRODUCTION

Diseases of the human heart represent one of the most important problems in medicine today. Computers are an indispensible tool in cardiac research and are used mainly in two research areas (*Figure 1*).

The first is modelling of the human heart. This research aims at gaining more insight into the functioning and structure of the human heart. It is obvious that modelling studies of the human heart could hardly be conducted without the aid of a computer.

The second is the handling of the cardiac data. This research aims at diagnosis, prognosis, and treatment. Although, the handling of some cardiac data can be performed manually, automating it by a computer has a lot of advantages and is worth the effort. The most important of these advantages are:
(i)   the data handling procedures are applied in a consistent manner
(ii) the results are reported in a standard format
(iii) it is possible to use a standard terminology.
The value of handling by computer cardiac data, which cannot be handled manually, needs no comment.

The various kinds of cardiac data mostly used in cardiac research today, are shown in *Figure 2*. The various tasks that computers perform in handling the cardiac data are the following five: acquisition, storage, retrieval, processing, and reporting. The processing task is further subdivided into the following subtasks: filtering, pattern recognition, parameter measurement, and interpretation.

In this chapter we will not cover modelling of the human heart and the handling of blood pressure and blood flow data. We will cover at some length the handling of electrocardiographic (ECG) data because the research is more mature in this area. We will also cover very briefly the handling of cardiac images.

2. ECG DATA HANDLING

The electrocardiogram (ECG) contains diagnostic information and for this reason it is widely used as a diagnostic tool in clinical practice. ECGs can be divided into three categories, depending on the purpose of their use. These categories are:
(i)   'rest ECGs' which are recorded from ambulatory patients at rest for the purpose of an easy and rapid assessment of the cardiac status of the patient as well as for monitoring the effect of treatment;
(ii) 'exercise ECGs' which are recorded from patients while performing a (predefined)

Computers in cardiac research

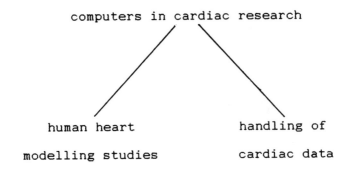

Figure 1. Uses of computers in cardiac research.

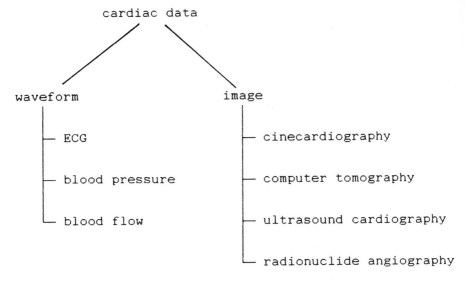

Figure 2. Categories of cardiac data.

exercise protocol for the purpose of (mainly) confirming or ruling out the presence of coronary artery disease;
(iii) 'arrhythmia monitoring ECGs' which are recorded either from critically ill patients in coronary and intensive care units or from ambulatory patients for the purpose of (mainly) detecting arrhythmias.

The process of extracting the diagnostic information from ECGs is called ECG analysis. Attempts towards automating this by computer started in the late fifties. Besides the benefits of automation, it was expected that the diagnostic accuracy of the analysis would increase by applying statistical methods which could not be applied manually. As a result various computer-based systems have been developed for ECG analysis. In these systems, ECG analysis is usually performed in four phases (1).
(i) ECG acquisition.
(ii) Signal conditioning.

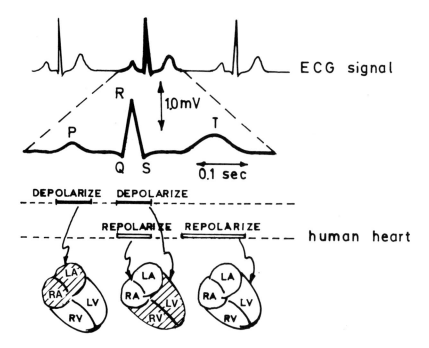

Figure 3. Correspondence between the ECG signal and the functioning of the human heart.

(iii) ECG recognition and parameter measurement.
(iv) ECG interpretation and classification.

In the rest of this section, the electrocardiographic patterns relevant to the ECG recognition and parameter measurement phase are first described and the parameters to be measured are identified. Each of the four phases is then analysed separately. More emphasis has been put on the pattern recognition problem because of its challenging nature. An illustrative example is also given together with the initial data in digital form to clarify the methods involved. The measurement of the parameters is performed at the time the patterns are recognized which is a straightforward task in view of the approach taken for the recognition of the ECG patterns.

2.1 Electrocardiographic patterns

The ECG is a set of waveforms, commonly known as leads. Each one of these waveforms is a recording of the electrical activity of the heart within a certain time period. The relation between the ECG signal and the functioning of the human heart is depicted in *Figure 3*. A cyclic pacemaking impulse normally originates in the sino-atrial (S-A) node, which is located high in the right atrium, and ends up at the ventricles. This impulse initiates a wave of cellular depolarization which is propagated throughout both atria and results in the contraction of the atria and the filling of the ventricles with blood. The impulse travels down to the atrio-ventricular (A-V) node where a delay is introduced before it is passed to the ventricles. This delay allows for completion of mechanical contraction of the atria. The impulse is then distributed via the bundle of His and Purkinje

Computers in cardiac research

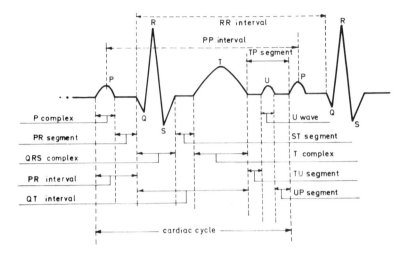

Figure 4. A cardiac cycle and its constituent patterns.

fibres to the more muscular ventricles. The arrival of the impulse in the ventricles initiates a wave of cellular depolarization which results in the contraction of the ventricles and the expelling of blood into the pulmonary and systemic circulations. After the contraction of the ventricles, a recovery period follows (depolarization), which lasts until the heart reaches its resting state. At this point, a cardiac cycle has been completed and a new cycle can be initiated through a new impulse generated at the S-A node etc. Thus, within a cardiac cycle we normally observe three complexes on the ECG waveform (see *Figure 4*). The P complex which corresponds to the depolarization of the atria, the QRS complex which corresponds to the depolarization of the ventricles and the T complex which corresponds to the repolarization of the ventricles. No visible complex corresponds to the repolarization of the atria.

Each cardiac complex consists of a series of peaks of alternating sign. The number of peaks in each complex varies from lead to lead and from patient to patient. The P complex and the T complex usually have one or two peaks. The QRS complex can have from one to seven peaks. The term 'wave' is used by physicians instead of the term peak. Both terms are used in the following account as equivalent.

The electrocardiographic patterns that must be recognized in the pattern recognition and pattern parameter measurement phase of an ECG processing system are the complexes, the interwave segments and the cardiac intervals (*Figure 4*). The parameters of these patterns that must be measured are height and duration for the complexes and duration for the interwave segments and the cardiac intervals. Thus, there are two types of measurements to be performed: 'time measurements' and 'amplitude measurements'. The time measurements are harder to obtain in practice. For the height measurements, the baseline is used as a reference.

2.2 ECG acquisition

The ECG is recorded by an instrument called the 'electrocardiograph'. Until the early sixties the ECG was recorded one lead after the other. Since the late sixties 3-channel

ECG recorders became available. With these instruments three ECG leads are recorded simultaneously. Recently, recording equipment has become available that records more leads simultaneously.

After the ECG has been recorded by the electrocardiograph it passes through an A/D (analogue-to-digital) converter. This is because the ECG waveforms, as they are recorded, are in analogue form and they have to be converted into digital form to be processable by a computer. Various sampling rates, such as 250, 300, 400, 500 or 1000 Hz, are used. The most commonly used are 250 and 500 Hz. Both are sufficiently high according to the Shannon theorem, since in the adult ECG the highest frequency which is considered to be of diagnostic interest, is approximately 80 Hz. For paediatric ECGs, however, a sampling rate of 250 Hz is clearly too low because of the higher frequency components in the ECGs. When a low sampling rate is used, interpolation between the sample points is sometimes performed in order to obtain reliable time measurements. With respect to storage needs, low sampling rates offer some advantage, although even then a considerable amount of data has to be stored.

The precision of the A/D conversion usually ranges from 8 to 12 bits. Each ECG waveform, after the digitization process, is represented in the computer as a series $y_1, y_2, \ldots, y_n$ where y_i is the amplitude in microvolts of the sample point i.

2.3 Signal conditioning

After an ECG signal has been recorded and sampled, digital filtering techniques are often applied to enhance its quality and suppress noise, which is mainly due to muscle tremor, electrode displacement and mains interference. The digital filters commonly used are low-pass designs (2). They fall generally into the three filter classes.

(i) *The moving average, or convolution filter.* This is the earliest and simplest type of digital filter. Each filter output $f(j)$ is calculated from a finite sequence of input samples $y(i)$ by an equation of the form

$$f(j) = \sum_{t=-p}^{p} w(t)y(j-t)$$

where $w(t)$ is a set of $2p + 1$ weighting coefficients. In the equation given it is implicit that an equal number of samples before and after $y(j)$, the point being smoothed are taken into account. If a further restriction is made, namely that the weighting coefficients are symmetrical about $y(j)$, that is

$$w(i) = w(-i)$$

then the phase response of this filter is 0 or π, and such a filter may loosely be referred to as a zero-phase filter.

(ii) *The recursive filter.* Not only input values but also previous output values are used as feedback to calculate the next output. For the iterative process to start, the initial output (or outputs) must be assumed. In general, for input $y(t)$, the output $f(j)$ is given by

$$f(j) = \sum_{t=1}^{p} u(t)f(j-t) + \sum_{t=m}^{q} v(t)y(j-t)$$

where the $u(t)$ and $v(t)$ are fixed coefficients. The value of p defines the order—in the sense of classical network analysis—of the filter. When $p=1$ for example, the filter is called a first-order recursive filter.

(iii) *The frequency-domain filter.* Filtering is performed in the frequency domain by use of the discrete Fourier transform (d.f.t.). The complex coefficients $S(j)$, which correspond to frequencies $0, 1/T, 2/T,...,(N-1)/T$ where T is the length of the transformed record, are calculated from the series $y(k)$:

$$S(j) = \frac{1}{N} \sum_{K=0}^{N-1} y(k) w^{kj}$$

$(w = e^{-2i\pi/N}, j=0,1,...,N-1)$

These coefficients are next scaled by multiplication with discrete values of the frequency response (amplitude) curve. Thereafter, inverse transforming reconstructs the filtered time signal from its suitably amplified or attenuated component harmonic sinusoids.

2.4 ECG recognition and parameter measurement

The task of ECG recognition is the hardest of all in any ECG processing system. In this phase the ECG patterns are recognized and their parameters are measured. When a signal conditioning phase has already taken place, the filtered signal is given as input to this phase. Otherwise, the original signal is given as input.

A bottom up method for the recognition of the ECG patterns is described here which is based on the concept of peaks. This method performs its task in three steps by calling the following procedures in order.

(i) The 'peak recognition' procedure which recognizes the real peaks (those not due to noise) in an ECG waveform.
(ii) The 'peak delineation' procedure which finds the boundaries of the recognized real peaks.
(iii) The 'ECG recognition' procedure which recognizes the ECG patterns using the peak patterns which have been extracted by the previous two procedures.

2.4.1 *Peak recognition procedure*

The term peak is used here in a descriptive rather than a strict mathematical sense. For a mathematical definition the reader is referred to reference 3. The peak pattern (*Figure 5*) is that part of a waveform which is demarcated by three characteristic points. The first point is called 'left peak boundary', the second 'peak extremum' and the third 'right peak boundary'. The sample points between the left peak boundary and the peak extremum form the 'left arm' of the peak. The sample points between the peak extremum and the right peak boundary form the 'right arm' of the peak. We will symbolize peaks as $P_1, P_2,...$, where P_i is the name of peak i.

A set of nine attributes is assigned to each peak pattern P_k. The values of these attributes are calculated during the peak recognition and delineation phases and they are utilized during the ECG recognition and parameter measurement phase. They contribute both to the recognition of the patterns and to the measurement of their parameters. That is, they are used in a quantitative way for qualitative and quantitative

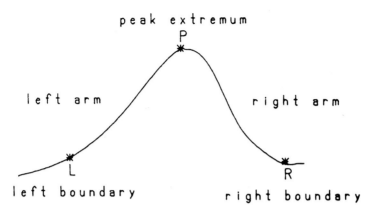

Figure 5. Illustration of the peak pattern.

purposes. This set is symbolized as

$$\{xl_k, yl_k, xm_k, ym_k, xr_k, yr_k, vl_k, vr_k, e_k\}$$

where: (xl_k, yl_k) is the left boundary of the peak P_k; (xm_k, ym_k) is the peak extremum of the peak P_k; (xr_k, yr_k) is the right boundary of the peak P_k; vl_k is the maximum slope at the left arm of the peak P_k; vr_k is the maximum slope at the right arm of the peak P_k; and e_k is the energy of the peak P_k defined as

$$e_k = \sum_{i=p}^{q} (y_i - y_{i-1})^2, \ p = xl_k + 1, \ q = xr_k$$

The peak recognition procedure performs its task in two stages (4). In the first stage it recognizes all peaks. Some of them are real peaks and some are noisy peaks due to noise. No distinction between the peaks is made in this stage.

In the first stage the recognition of all the peaks is accomplished by detecting changes in the sign of the slope. That is, the sample point y_i is considered to be a peak if $(y_i - y_{i-1}) \times (y_{i+1} - y_i) < 0$. If the difference $y_{i+1} - y_i$ is equal to zero (a flat portion in the waveform), then it is ignored and instead the difference $y_{i+2} - y_{i+1}$ is considered and so on until the second term in the multiplication is not equal to zero.

In the second stage the noisy peaks are recognized and subtracted from the set of all peaks recognized in the first stage and the remaining peaks are the real ones. This is accomplished in four steps as described below.

(i) *Recognition of noisy peak pairs.* In this step noisy peak pairs are recognized and rejected. A noisy peak pair is a pair of consecutive peaks that satisfy a set of criteria. When the peak P_i is positive, the peak pair (P_{i-2}, P_{i-1}) is considered as noisy (*Figure 6*), according to the following criteria:

criterion 1: $yl_{i-2} \leq yl_{i-1}$ and $yr_{i-1} \geq yr_{i-2}$

criterion 2: $\Delta t_m \leq \epsilon$ where $\epsilon = \epsilon_2$ when $|ym_{i-1} - ym_{i-2}| < \epsilon_1$

else $\epsilon = \epsilon_3$

criterion 3: $\Delta t_m \geq \Delta t_l$ and $\Delta t_m \geq \Delta t_r$ when $\Delta t_m > \epsilon_4$

Computers in cardiac research

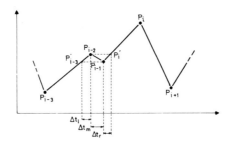

Figure 6. Noisy peak pair.

When the peak P_i is negative the inequalities in the first criterion are reversed while the other two criteria are the same. ϵ_1, ϵ_2, ϵ_3 and ϵ_4 denote threshold values.

The first of these criteria demands that the peak pair (P_{i-2}, P_{i-1}) is located (nested) within the peak pair (P_{i-3}, P_i). The second and third criteria impose some duration constraints.

It has been found (5) that, on average, 84% of the noisy peaks in an ECG waveform can be recognized this way. These peaks are rejected but in the remaining set some noisy peak pairs are still left which are mainly located near the baseline. These noisy pairs are recognized by the following criteria which are applied to each peak pair (P_i, P_{i+1}):

criterion 4: $xm_{i+1} - xm_i < \epsilon_5$

criterion 5: $|ym_{i+1} - ym_i| < \epsilon_6$

criterion 6: $|yl_i - ym_i| < \epsilon_7$

criterion 7: $|yr_{i+1} - ym_{i+1}| < \epsilon_8$

The first is a duration criterion while the others are height criteria. The above criteria state that a pair of adjacent peaks which have small duration and amplitude are rejected as noisy. ϵ_5, ϵ_6, ϵ_7 and ϵ_8 denote threshold values.

(ii) *Recognition of noisy peaks near the baseline.* In this step noisy peaks near the baseline are recognized and rejected. For this reason an approximation to the baseline is first calculated and then the noisy peaks near it are recognized and rejected. An approximation to the baseline is calculated using the set of noisy peaks recognized as noisy peak pairs mainly located near the baseline in the previous step. Thus, the peaks referred to in this step are from this set. The calculation of an approximation to the baseline is based on the assumption that (i) the variance of each arm in small peaks close to the baseline is small (*Figure 7a*) and (ii) the variance of one of the arms, on peaks close to the baseline as in *Figure 7b* and *Figure 7c*, is also small.

Such peaks which are close to the baseline are identified and used for finding fiducial points from which the approximation to the baseline passes. To each such peak P_i, a fiducial point j is associated with coordinates (xb_j, yb_j) where $xb_j = xm_i$ and yb_j is defined below. The approximation to the baseline is denoted by the ordered set of fiducial points

$$\{(xb_1, yb_1), (xb_2, yb_2), \ldots, (xb_n, yb_n)\}$$

where n is the number of fiducial points found. The actual approximation to the baseline

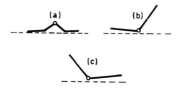

Figure 7. Peaks close to the baseline that can be used for finding fiducial points.

is formed by joining these fiducial points by line segments.

For identifying that the peak P_i can be used to find a fiducial point the following criterion is used:

$$\text{criterion 8: } Z = Z_1 \text{ or } Z_2$$

where Z is a predicate variable; Z_1 and Z_2 are predicate variables with values given by the formulas

$$Z_1 = (\sigma_l)^2 \leq \epsilon_9$$
$$Z_2 = (\sigma_r)^2 \leq \epsilon_9$$

$$(\sigma_l)^2 = \sum_{k=b}^{t} (y_k - \bar{y}_l)^2, \ t = xm_i, \ b = t - \epsilon_{10}, \ \bar{y}_l = (\sum_{k=b}^{t} y_k)/(\epsilon_{10}+1)$$

$$(\sigma_r)^2 = \sum_{k=b}^{t} (y_k - \bar{y}_r)^2, \ b = xm_i, \ t = b + \epsilon_{10}, \ \bar{y}_r = (\sum_{k=b}^{t} y_k)/(\epsilon_{10}+1)$$

$\epsilon_9, \epsilon_{10}$ are threshold values.

A variable q is defined according to: $q=1$ when both Z_1 and Z_2 are true, $q=2$ when only Z_1 is true and $q=3$ when only Z_2 is true. The yb_j coordinate of the corresponding fiducial point is calculated according to

$$yb_j = (\bar{y}_l + \bar{y}_r)/2 \quad \text{when } q=1$$
$$yb_j = \bar{y}_l \quad \text{when } q=2$$
$$yb_j = \bar{y}_r \quad \text{when } q=3$$

The recognition and rejection of noisy peaks near the baseline is accomplished by applying the following criteria to each peak P_i.

$$\text{criterion 9: } \neg Z$$
$$\text{criterion 10: } xr_i - xl_i < \epsilon_{11}$$
$$\text{criterion 11: } |ym_i - yb_i| < \epsilon_{12}$$

where Z, Z_1, Z_2 are predicate variables with values given by the formulas:

$$Z_1 = |ym_i - yl_i| > \epsilon_{13} \text{ and } |ym_i - yr_i| > \epsilon_{14}$$
$$Z_2 = |ym_i - yl_i| > \epsilon_{14} \text{ and } |ym_i - yr_i| > \epsilon_{13}$$
$$Z = Z_1 \text{ or } Z_2$$

where yh_i is the amplitude of the baseline (found from its approximation) at the point with x-coordinate equal to the one of peak P_i. $\epsilon_{11}, \epsilon_{12}, \epsilon_{13}$ and ϵ_{14} are threshold values.

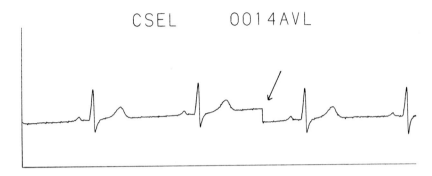

Figure 8. Illustration of a jump.

The criteria have the following meaning. Criterion 9 protects small Q and S waves from rejection as noisy peaks. The need for this criterion comes from the fact that these waves are often close to the baseline. Criterion 10 demands that noisy peaks have small duration and criterion 11 states that peaks close to the baseline are considered as noisy.

The peaks recognized as noisy are subtracted from the input set of peaks given to this step and the set of the remaining peaks is given as input to the next step.

(iii) *Recognition of jumps.* In this step jumps are recognized and rejected. As illustrated in *Figure 8*, a jump is a noisy peak (an artefact). In a jump there is a sudden and large deviation from the baselevel that does not return to it within a short time interval. The net effect of this artefact is that the baseline changes level. A peak P_k is recognized as a jump when it satisfies the following two criteria:

criterion 12: $|vl_k| > \epsilon_{15}*|vr_k|$ or $|vr_k| > \epsilon_{15}*|vl_k|$

criterion 13: max $\{|ym_k-yl_k|, |ym_k-yr_k|\} > \epsilon_{16}$

where ϵ_{15}, ϵ_{16} are threshold values.

Criterion 12 expresses the property that the maximum slope in one arm of a jump is considerably greater than the maximum slope of the other arm. Criterion 13 expresses the high amplitude characteristics of the jumps.

The peaks recognized as jumps are subtracted from the input set of peaks given to this step and the set of the remaining peaks is given as input to the next step.

(iv) *Recognition of high spikes.* In this step high spikes are recognized and rejected. As illustrated in *Figure 9*, a high spike is a sequence of consecutive noisy peaks (an artefact).

In a high spike there is a sudden large deviation from the baselevel that returns to it within a short time interval. The characteristics of a high spike are steep slopes, high amplitudes and small duration. A sequence of consecutive peaks is recognized as a high spike when the following criteria are satisfied by each peak:

criterion 14: $|ym_k-yl_k| \leq e_{17}*|vl_k|$ or $|ym_k-yr_k| \leq \epsilon_{17}*|vr_k|$

criterion 15: min $\{|ym_k-yl_k|, |ym_k-yr_k|\} > \epsilon_{18}$

criterion 16: $xr_k-xl_k < \epsilon_{19}$

where ϵ_{17}, ϵ_{18}, ϵ_{19} are threshold values.

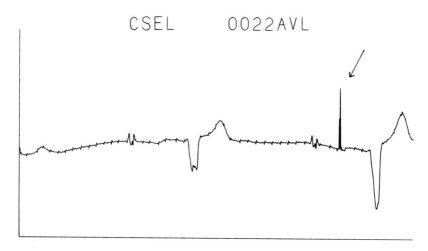

Figure 9. Illustration of a spike.

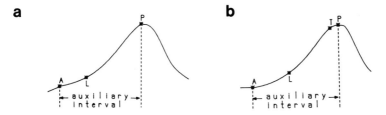

Figure 10. Illustration of the maximum curvature criterion.

Criterion 14 expresses the steep slope characteristic, criterion 15 the high amplitude characteristic and criterion 16 the small duration characteristic.

The peaks recognized as high spikes are subtracted from the input set of peaks given to this step and the set of the remaining peaks is the set of the real peaks. This set is given as input to the peak delineation procedure.

2.4.2 Peak delineation procedure

A precise method for peak delineation in ECG waveforms is the 'maximum curvature criterion' method. This method is based on the assumption that the boundary points of a peak—onset and offset—are points where the curvature takes its maximum value over an interval around that points. This is illustrated in *Figure 10a* where the interval mentioned is called 'auxiliary interval'. For a peak morphology as the one shown in *Figure 10b* this assumption may not be valid because the curvature at points T and L is comparable.

Properties of the second derivative of the curve can be used to overcome this problem. That is, the sign of the second derivative, which at point T for example (convex) is minus whereas at point L (concave) is plus can be taken into account. So the above assumption can be stated as follows: at a boundary point of a positive (negative) peak the curvature is maximum over an interval which contains that point, considering only

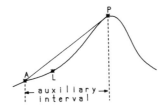

Figure 11. Auxiliary interval definition.

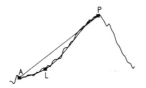

Figure 12. Curve approximation by cubic spline fitting superimposed on the actual data points.

the values of the curvature at the points where the second derivative is positive (negative).

The auxiliary interval, which is the search interval for a peak boundary point, is defined by a piecewise linear approximation (PLA) method (6). Starting from the point P, a number of data points are approximated by a straight line segment while the approximation error is required to be less than a preset threshold. The point A (*Figure 11*) defined in this way, is the end point of the longest straight line segment which satisfies the above constraint. If the approximation error is permited to be large enough, then one can be sure that the real peak boundary is contained in the auxiliary interval. Experimentally, a threshold value for the approximation error, such that the point A has the desired property, can very easily be found and furthermore this value is not very critical.

For the computation of the curvature K_t at any point t of a curve $y=y_t$ the following formula is used

$$K_t = |y_t''|/[1+(y_t')^2]^{3/2}$$

This formula can be computed at any point t, provided that the first and second derivatives of the curve y_t can be computed. Since ECG waveforms are not smooth due to noise, the derivatives of the curve are discontinuous (or do not exist) at some points and in this case the curvature can not be computed. Thus, the curve y_t has to be approximated within the auxiliary interval by a smooth function with continuous first and second derivatives. Curve approximation can be performed by fitting a function to the original data points. The cubic spline function is a suitable function for the case of ECG waveforms. This function consists of a number of cubic polynomial segments joined end to end, with continuity in first and second derivatives at the joins. In our experiments we have used routines from the NAG FORTRAN library for approximating a curve segment, since it offers a variety of routines for the computation of the cubic spline and its derivatives.

Curve approximation is illustrated in *Figure 12*, where the results of the cubic spline approximation are superimposed on the actual data points. The point L shown in this figure, is the peak boundary detected by the maximum curvature criterion.

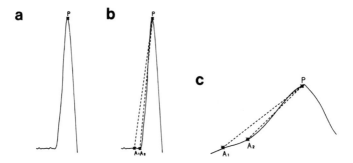

Figure 13. (a) Illustration of a peak with arms very close to straight lines. (b) The results of the PLA algorithm applied twice in the peak of *Figure 11a*. (c) the results of the PLA algorithm applied twice in the peak of *Figure 10*.

(i) *Method refinement*. For some peaks, such as the one shown in *Figure 13a*, the boundary points are often erroneously detected with this method. This is mainly due to poor approximation of the data points by the cubic spline function. It is noted that the arm of such a peak is a straight line and its boundary can easily be computed by a PLA algorithm. To overcome this problem, peak arms like the one mentioned above are detected prior to the fitting of the cubic spline function. This is accomplished by applying the PLA algorithm twice, the first time permitting a large error value in the approximation, the second time a small one. If the two points found are close enough (*Figure 13b*), no cubic spline fitting is applied and the point found by the second application of the PLA algorithm (point A_2) is accepted as the boundary point. In the opposite case (*Figure 13c*), the maximum curvature criterion method is normally applied, with the auxiliary interval defined by the point found by the first application of the PLA algorithm (point A_1) and the peak extremum.

2.4.3 ECG recognition procedure

The recognition of the electrocardiographic patterns is considered here. We will focus on the recognition of the cardiac complexes only, since the other patterns—interwave segments and cardiac intervals—are readily available when the cardiac complexes have been recognized. The QRS complexes, as the most prominent patterns in ECG waveforms, are recognized first. Classification of these complexes into morphological classes follows. For the recognition of the P complexes and the T complexes, *a priori* structural relationships and information carried by the peak attributes is used.

(i) *Recognition of QRS complexes*. Recognition of the QRS complexes is performed in two steps as follows.

(a) In this step, peaks are combined to form 'peak complexes'. A series of 'consecutive peaks' (peaks without intervening segments) is considered as a peak complex when, in each pair of consecutive peaks, the angle between the right arm of the first peak of the pair and the left arm of the second peak of the pair, is less than ϵ_{20}, where ϵ_{20} is a threshold value. To each peak complex an energy attribute is assigned which is calculated as the sum of the energies of its constituent peaks.

(b) In this step the QRS complexes are recognized. A peak complex is recognized as a QRS complex when its energy is greater than $\epsilon_{21}*e_{max}$, where ϵ_{21} is a threshold value and $e_{max}=max\ e_k$. In other words, the decision function, for k characterizing a peak complex as a QRS complex, is thresholding. This decision function can be justified as follows: visually, the QRS complexes consist of a series of consecutive abrupt peaks, that is the qualitative feature that differentiates the QRS complexes from the other complexes is the abruptness of their constituent parts (peaks). This feature quantitatively can be expressed by the energy attribute. That is, the energy attribute is considerably greater in QRS complexes than in any other complexes. Once a peak complex has been characterized as QRS complex, the recognition of its morphology is a trivial task. Indeed, a QRS complex can have from one up to seven peaks of alternating sign. The naming of these peaks is performed in a table-lookup manner. The entries in the table are of the form:

peak complex morphology, QRS morphology

These entries are shown in *Table 1* where $P+$ stands for positive peak and $P-$ stands for negative peak. It is noted that the onset and offset as well as other measurements of the QRS complexes are readily available from the attributes of the constituent peaks.

(ii) *Classification of QRS complexes.* QRS classification involves the separation of the QRS complexes in a given ECG waveform into classes based on their degree of morphologic similarity. It is an essential process for arrhythmia analysis in computerized systems.

QRS classification is performed by a 'feature extraction/classification' method (7). In this method, the QRS data are represented by a set of either formal features such as coefficients of orthonormal vector sets or heuristic features which have problem related meanings. Heuristic features offer the advantage of extremely efficient computations in terms of time. Such features commonly used are QRS amplitude, width, centre of mass offset from the baseline, area, time interval between peaks etc., which are measured after a QRS complex has been recognized. Thus, individual QRSs are represented as a series of numbers which are stored in the elements of a 'data vector'.

Table 1. QRS morphology recognition.

Peak complex morphology	QRS morphology
$P+$	R
$P-$	QS
$P-P+$	QR
$P+P-$	RS
$P-P+P-$	QRS
$P+P-P+$	RSR'
$P-P+P-P+$	QRSR'
$P+P-P+P-$	RSR'S'
$P-P+P-P+P-$	QRSR'S'
$P+P-P+P-P+$	RSR'S'R''
$P-P+P-P+P-P+$	QRSR'S'R''
$P+P-P+P-P+P-$	RSR'S'R''S''
$P-P+P-P+P-P+P-$	QRSR'S'R''S''

Once the data vectors are defined, the set of QRSs, symbolized as $\{QRS_1, QRS_2, \ldots, QRS_n\}$ is partitioned into m QRS classes $\{C_1, C_2, \ldots, C_m\}$ by a nearest neighbour classification algorithm. This algorithm works in three steps as follows:
(1)   Assign QRS_1 to C_1. $j=1$, $m=1$.
(2)   Increase j by 1.
Compute $D = \min d(C_1, QRS_j)$, where $d(C_1, QRS_j)$ stands for the distance between class C_1 and QRS_j. Assign to i the value 1 for which the above distance is minimum. If D is less than or equal to t then assign QRS_j to C_i else initiate a new class for QRS_j and increase m by 1.
(3)   Repeat step 2 until all the QRSs have been assigned to classes.

The parameter t used in this algorithm is a threshold value needed for class separation. The distance d between a class of QRS complexes and a QRS complex is computed as the average of the distances between the QRS complex and each QRS complex in the class. Since the QRS complexes are represented as a vector in a N-dimensional space (N is the number of features used), the distance between two QRSs is expressed as either the 'straight line' (L2) or the 'city block' (L1) distance between N-space locations. For two vectors X and Y these distances are defined as:

$$L1: D = \sum_{i=1}^{N} |X_i - Y_i|$$

$$L2: D = \sum_{i=1}^{N} (X_i - Y_i)^2$$

The L1 distance gives better results and it is simpler to compute. Some form of normalization, such as 'magnitude normalization', is applied to the distance computations in order to make them more universal and independent of scaling factors. In the magnitude normalization the similarity measures are divided by the magnitudes of the vector elements of one or both QRS beats. The class that has the maximum number of QRSs is considered to be the representative class and the QRSs in it are called representative QRS complexes.

(iii) *Recognition of P and T complexes.* P and T complexes are recognized only in cycles with representative QRS complexes. A search for a P or a T complex is preformed in a region before (after) a representative QRS complex. One or two consecutive peaks in this region are recognized as a P (T) complex, by thresholding their width and amplitude (thresholds ϵ_{22} and ϵ_{23}, respectively). For simplicity, P and T complexes before the first QRS complex found and after the last QRS complex found are not recognized. The recognition of a P or T complex morphology is performed in a table-

Table 2. P (T) morphology recognition.

Peak complex morphology	P (T) morphology
$P+$	P (T)
$P-$	P$-$ (T$-$)
$P-P+$	P$-$P (T$-$T)
$P+P-$	PP$-$ (TT$-$)

Computers in cardiac research

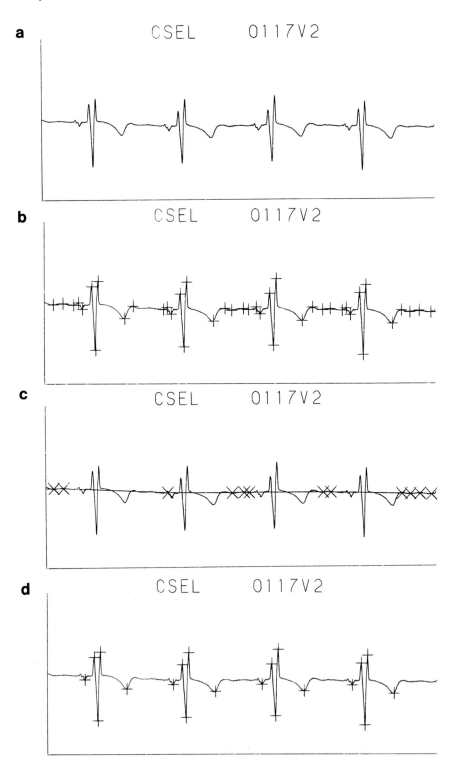

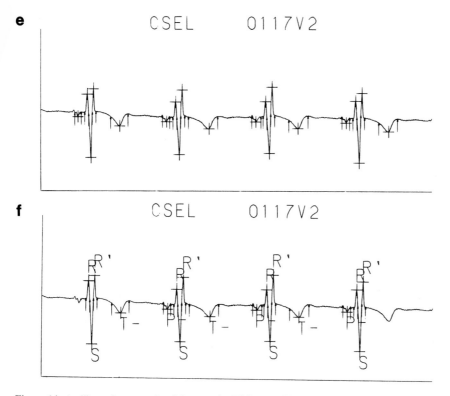

Figure 14. An illustrative example of the steps in ECG recognition.

lookup manner similar to the one used for QRS complexes. The table entries (which are the same for P or T complexes) are shown in *Table 2*.

2.5 An illustrative example

An illustrative example is given in *Figure 14* where the results of the various steps in the recognition procedure of the electrocardiographic patterns are shown. The initial ECG waveform is given in *Figure 14a*. The sample points of this waveform (sampled at the rate of 500 samples per second) are given in Appendix *Listing 1*. This sample ECG waveform has not been filtered. This is because the noise level of the library containing this ECG (8) is within acceptable limits. Thus, the original ECG waveform has been given as input to the peak recognition procedure. The results of the various steps of this procedure are given in *Figure 14b−d*. The set of noisy peaks recognized as noisy peak pairs were subtracted from the set of all peaks recognized in stage 1 and the remaining peaks are shown in *Figure 14b*, where the peaks are marked by the symbol '+'. The baseline approximation calculated is shown in *Figure 14c*, where the fiducial points are marked by the symbol 'x'. The noisy peaks near the baseline were subtracted from the set of peaks of *Figure 14b* and the remaining peaks are shown in *Figure 14d*. No noisy peaks were recognized as high spikes or jumps.

The results of the peak delineation procedure are given in *Figure 14e* where the calculated peak boundaries are marked by up arrows (↑).

The results of the ECG recognition procedure are given in *Figure 14f* where the names of the recognized ECG waves have also been plotted. QRS classification resulted in the formation of one class containing all the QRS complexes of this lead.
The values for the thresholds used are: $\epsilon_1=25$, $\epsilon_2=16$, $\epsilon_3=20$, $\epsilon_4=4$, $\epsilon_5=15$, $\epsilon_6=65$, $\epsilon_7=\epsilon_8=200$, $\epsilon_9=5$, $\epsilon_{10}=30$, $\epsilon_{11}=8$, $\epsilon_{12}=50$, $\epsilon_{13}=200$, $\epsilon_{14}=20$, $\epsilon_{15}=6$, $\epsilon_{16}=250$, $\epsilon_{17}=2$, $\epsilon_{18}=250$, $\epsilon_{19}=15$, $\epsilon_{20}=30$, $\epsilon_{21}=0.3$, $\epsilon_{22}=10$, $\epsilon_{23}=20$.

2.6 ECG interpretation and classification

The results of the ECG pattern recognition and pattern parameter measurement phase are used in the interpretation phase to arrive at diagnostic classification statements. Two major approaches are used in this task.

The first and most widely used is a deterministic approach which tries to automate the clinical procedure used by the cardiologist (9). In this approach, binary decisions which test the wave measurements against established criteria lead to a set of non-conflicting statements about the ECG. It is noted that no set of criteria exists today that is universally acceptable. The establishment of such a set is a research matter.

In the second approach, multivariate statistical classification techniques are applied to calculate the probability of a particular classification (10). In this approach, initial patient classification in diagnostic categories is based on independent, that is, non-ECG information (i.e. cardiac catheterization, coronary angiography, autopsy results, etc.). Probability calculation is then based on a large number of ECG variables which are used simultaneously, usually in the form of multidimensional vectors. General statistical properties of the measurements and Bayes procedures are used with the aim of maximizing correct classification and minimizing incorrect classification of the ECG.

3. CARDIAC IMAGE HANDLING

So far, we have discussed computer analysis of ECG data. These data are acquired in the form of sampled waveforms, each of which is adequately represented by a few thousand bytes. Thus the storage requirements per patient, although large, are nominal. However, this is not the case with the data generated by medical images. Storage requirements for digital (medical) images are very large and in general big—or special purpose image analysing—computers are needed for image data processing.

3.1 Cardiac imaging techniques

The standard method, traditionally used for cardiac diagnosis, is cardiac catheterization with intravascular pressure recording and angiocardiography. Catheterization involves the threading of a thin plastic tube from an artery in the leg through the arterial system and into the coronary arteries of the heart where pressure readings are made and radio-opaque dye can be injected. Angiocardiography represents a method of X-ray fluoroscopic imaging of the heart that is performed during catheterization. X-ray fluoroscopic imaging alone does not show the structures of the heart or the anatomy of the coronary artery tree that supplies blood to the heart muscle. This can be achieved by using direct injections of radio-opaque contrast medium directly into the structures of interest via a catheter. Such invasive procedures are not preferred and thus a number of alternative imaging techniques have been developed for non-invasive cardiac imaging (11,12).

Nuclear medicine techniques have been impressively developed during the last few years for the evaluation of the heart. Digital computers are used to form images generated by the radiation from radionuclides. Due to the low resolution power of nuclear imaging, sophisticated image preprocessing techniques, such as smoothing and image enhancement, are required.

Ultrasonic scanning of the heart, also known as echocardiography, yields higher spatial resolution than nuclear medical imaging. In recent years, several systems capable of two-dimensional ultrasonic visualization of the heart in real time have been developed. Information is entered in the computer with sonic digitizers or by the conversion of transparencies to digital data.

Computed X-ray tomography (CT) is a very sensitive method for heart imaging. The advantages of CT over other imaging methods are derived from the fact that very small density variations can be adequately detected and spatially located. Computed X-ray tomography systems are one of the few in which the computer is a vital part of the overall installation. Normally a 16- or 32-bit minicomputer is used, together with one or more floating point processors, to form the image from the raw projection data through standard convolution and back-projection methods. In this case, a cross-sectional image is reconstructed from the many one-dimensional projections of the same cross-section.

3.2 Cardiac image processing

Determination of the contours of the left ventricle is of high importance for diagnostic purposes. Thus, after an image of the heart has been constructed, digital image processing is often applied for the detection of the left ventricle contours (13). The image is represented in the computer as a non-negative function (matrix) $f(x,y)$ of two variables. This function has to be proportional to the light intensity emerging from a picture at the picture element (pixel) (x,y). Processing is often performed in two stages: pre-processing, that is, filtering or smoothing and edge detection.

A fast and commonly used pre-processing method is a nine-point smoothing algorithm. The nine pixel values of a 3 × 3 window are averaged with different weighting coefficients, depending on the importance given to the central pixel or to its neighbours, and the central pixel is replaced by the average value. This filter performs an alternation of higher intensity variation in a given region, but also has the side-effect of blurring the 'sharpness' of ventricular boundaries.

In the median filtering pre-processing method, an ascending order classification in a given window is performed, and the considered pixel is filled with the median value. Noise is thus eliminated and the boundaries are not damaged but it has been experimentally observed that automatic edge detection, with this pre-processing method, is not noticeably improved.

Another pre-processing method is variant filtering; considering four windows around a central pixel, mean and variance of each window are computed. The value of the central pixel is replaced by the mean value of the window having the minimal variance. The performance of this filter is very good and this stems from the fact that the signal-to-noise ratio is related to the mean to variance ratio. Considering the mean value corresponding to the smallest variance higher signal-to-noise ratio is achieved on one hand, and on the other averaging on transition-region-like boundaries is avoided. Contrast between the object and background is thus increased and the contour sharpness is preserved.

Following the pre-processing phase, an edge operator is usually applied to facilitate ventricular edge detection. Several edge operators, derived from gradient techniques, have been proposed. An efficient one is the Sobel operator defined by the following mask, not taking central pixel into account

$$\begin{matrix} 1 & 0 & -1 \\ 2 & 0 & -2 \\ 1 & 0 & -1 \end{matrix}$$

This operator is not sensitive to noise and it is very efficient in cases of images with poor signal-to-noise-ratio. Moreover, its performance can further be improved when a variant filter pre-processor is used.

Boundary detection is performed on the resulting image. Local maxima detection, by thresholding, is a commonly used method. Another approach is based on fuzzy sets theory. In this approach a class of objects (pixels) is described with no precise belonging criteria but with belonging ratios, and a choice of isocontours gives the extracted boundaries of the left ventricle.

4. REFERENCES

1. Willems,J.L. (1981) *Endeavour, New Series*, **5**, 37.
2. Taylor,T.P. and Macfarlane,P.W. (1974) *Med. Biomed. Eng.*, p. 493.
3. Horowitz,S.L. (1977) *In Syntactic Pattern Recognition Applications*. Fu,K.S (ed.), Springer Verlag, Berlin,FRG, p. 31.
4. Skordalakis,E. and Trahanias,P. (1986) *8th IAPR Conf., IEEE Comp. Soc.*, p. 380.
5. Skordalakis,E. (1984) *Comput. Biomed. Res.*, **17**, 208.
6. Pavlidis,T. (1971) *IEEE Trans. Comput.*, **C-20**, 59.
7. Rappaport,S.H., Gillick,L., Moody,G.B. and Mark,R.G. (1982) *Computers in Cardiology*. IEEE Press, p. 33.
8. Willems,J.L. (1985) *Comput. Biomed. Res.*, **18**, 439.
9. Wartak,J., Milliken,J.A. and Karchmar,J. (1970) *Comput. Biomed. Res.*, **4**, 344.
10. Pipberger, H.V. (1975) *Am. J. Cardiol.*, **35**, 597.
11. Boyd,D.P. and Lipton,M.J. (1983) *Proc. IEEE*, **71**, 298.
12. Preston,K., Fagan,L.M., Huang,H.K. and Pryor,T.A. (1984) *Computer*, p. 294.
13. Romary,D., Lerallut.J.F. and Fontenier,G. (1985) *Comput. Biomed. Res.*, **18**, 488.

APPENDIX
Listing 1. Initial data sample points of the lead.

146	146	146	136	136	141	146	136	136	136	131	126	117	121	126	126	121
112	117	117	117	112	107	107	107	107	107	97	97	92	92	97	97	92
97	107	112	107	97	92	97	97	97	97	102	102	97	97	97	97	97
97	102	107	107	107	112	107	107	107	102	102	102	102	107	107	107	107
107	107	107	112	112	117	117	112	97	97	112	112	107	107	117	117	117
112	112	107	107	107	107	107	107	107	107	112	112	107	107	97	97	102
102	102	107	107	107	97	97	92	97	107	97	92	97	97	97	92	92
92	92	97	92	92	97	97	92	92	87	87	92	87	82	82	92	97
97	92	92	92	97	97	97	92	97	97	97	92	92	97	117	126	131
131	136	136	121	97	78	68	73	73	78	78	87	92	92	78	58	34
14	9	0	-10	0	19	34	29	34	43	73	92	73	58	73	82	92
97	97	92	87	87	87	87	92	92	87	82	82	82	87	92	92	92
92	97	92	92	97	117	165	239	312	370	429	468	487	482	443	365	273
175	58	-20	-79	-166	-283	-381	-449	-513	-649	-820	-903	-908	-825	-669	-469	-264
-79	92	224	-268	307	424	546	604	575	453	307	209	156	131	117	107	112
107	107	107	107	102	97	102	102	97	102	102	107	97	97	97	97	97
97	97	97	97	97	97	97	92	92	92	92	87	87	87	87	82	78
78	73	78	78	73	73	73	73	68	68	63	68	68	63	53	58	58
58	53	48	43	39	39	39	34	34	34	34	24	19	14	14	14	9
9	4	-5	-10	-20	-20	-20	-25	-30	-30	-35	-40	-40	-44	-49	-54	-64
-69	-79	-79	-88	-93	-98	-98	-108	-122	-127	-137	-142	-147	-157	-161	-166	-171
-176	-186	-191	-196	-201	-205	-215	-205	-210	-210	-215	-210	-201	-196	-186	-181	-176
-157	-142	-127	-118	-113	-98	-79	-69	-59	-44	-35	-25	-15	-10	0	4	14
19	29	29	34	39	43	43	39	39	48	53	53	53	53	53	58	58
53	53	53	53	53	48	48	43	39	39	39	34	34	34	29	24	29
29	29	19	19	19	19	19	19	14	19	19	19	14	14	14	14	19
14	14	14	14	14	14	14	14	14	14	14	14	19	14	14	14	14
14	14	19	19	14	14	14	19	14	14	14	14	19	19	19	19	19
29	19	19	14	19	24	29	29	34	29	24	19	19	14	19	14	14
14	19	24	29	19	14	9	14	14	14	9	14	14	19	14	4	-5
-5	4	9	4	0	4	9	4	0	0	4	0	-5	-5	0	4	4
0	-10	-10	-5	0	-5	-5	0	0	0	-10	-15	-15	-10	-10	-10	-10
-10	-10	-10	-20	-10	0	0	0	0	0	0	-5	-5	-10	-15	-10	0
9	19	19	29	29	19	4	-15	-40	-44	-49	-54	-54	-49	-44	-40	-49
-64	-93	-113	-118	-113	-113	-118	-113	-88	-79	-74	-69	-64	-54	-44	-40	-30
-20	-5	0	0	-10	-20	-20	-20	-20	-20	-10	-5	-10	-20	-20	-20	-15
-15	-15	-10	-10	0	0	0	9	43	102	165	214	273	312	321	302	263
180	82	-15	-122	-215	-254	-298	-361	-440	-508	-581	-659	-747	-820	-840	-801	-713
-566	-376	-166	14	170	297	341	365	453	546	585	565	453	292	175	102	53
29	29	29	29	19	14	14	19	14	9	4	9	4	0	-5	-5	-10
-10	-15	-10	-10	-5	-10	-10	-10	-15	-20	-20	-20	-20	-20	-20	-20	-25
-30	-35	-35	-35	-30	-35	-40	-40	-35	-30	-40	-49	-49	-49	-49	-49	-54
-54	-59	-64	-59	-59	-64	-74	-74	-74	-79	-79	-79	-79	-88	-93	-93	-88
-88	-93	-98	-98	-98	-103	-108	-118	-118	-118	-118	-127	-132	-127	-132	-137	-147
-157	-157	-161	-166	-171	-176	-176	-196	-205	-205	-205	-205	-215	-220	-230	-235	-240
-240	-240	-244	-254	-259	-274	-274	-269	-274	-274	-274	-274	-274	-274	-274	-269	-264
-254	-249	-235	-225	-220	-215	-205	-186	-171	-157	-147	-137	-118	-108	-98	-88	-88
-79	-74	-59	-48	-44	-35	-35	-25	-20	-25	-25	-20	-20	-10	-5	-5	-5
-5	0	0	0	-5	0	0	0	4	4	0	0	0	0	0	4	4
0	0	0	0	0	-5	-5	0	0	-10	-10	-5	-10	-15	-15	-20	-25
-20	-25	-25	-20	-15	-25	-30	-25	-25	-25	-30	-30	-35	-40	-35	-25	-20
-20	-30	-30	-25	-20	-20	-25	-30	-20	-10	-20	-30	-20	-10	-20	-25	-20

Computers in cardiac research

-10	-5	-10	-10	-10	-5	-10	-20	-15	-10	-10	-10	-10	-10	-10	-10
-15	-10	-5	-5	-5	-5	-5	-5	-5	-10	-10	-10	-15	-10	-5	0	0
-5	-10	-10	-10	-10	-15	-10	-10	-5	-5	-15	-20	-15	-15	-15	-15	-10
-10	-10	-20	-25	-25	-25	-20	-15	-20	-20	-20	-20	-20	-20	-20	-20	-20
-15	-15	-15	-10	-10	-10	-15	-20	-20	-20	-5	0	0	0	0	-5	-20
-15	-10	-10	-10	0	14	29	34	29	19	9	-15	-40	-49	-59	-64	-59
-59	-54	-59	-59	-69	-98	-118	-118	-108	-108	-108	-103	-93	-79	-69	-69	-64
-54	-40	-35	-25	-20	-20	-10	0	0	0	-10	-15	-10	0	0	-5	0
0	4	4	0	0	9	14	9	0	4	14	29	68	121	190	248	297
331	331	317	282	200	87	-15	-118	-201	-240	-279	-352	-435	-498	-557	-635	-737
-810	-810	-766	-679	-527	-322	-118	58	224	346	395	443	546	629	653	604	473
307	185	112	73	53	48	43	34	34	29	29	24	29	29	29	19
14	14	19	14	14	14	14	14	14	14	4	4	4	4	4	0	4
4	4	4	-5	-10	-10	-10	-15	-20	-20	-20	-20	-20	-30	-30	-30	-30
-30	-30	-30	-30	-30	-35	-40	-49	-49	-49	-49	-49	-59	-59	-59	-54	-59
-64	-69	-69	-69	-69	-74	-79	-79	-79	-88	-98	-98	-98	-98	-108	-108	-108
-113	-118	-127	-137	-142	-147	-152	-157	-157	-157	-166	-176	-181	-186	-196	-201	-201
-210	-215	-215	-220	-225	-235	-240	-244	-244	-254	-254	-259	-259	-264	-264	-259	-259
-259	-259	-254	-244	-240	-225	-220	-205	-196	-181	-176	-157	-147	-137	-122	-108	-98
-83	-79	-64	-59	-59	-44	-35	-25	-20	-20	-15	-10	-10	-5	0	9	9
9	14	19	24	14	14	14	14	19	19	24	29	24	19	14	14	19
19	19	14	14	19	14	14	4	4	0	0	0	0	0	0	-5	-10
-10	-15	-20	-20	-20	-20	-15	-15	-20	-20	-25	-25	-30	-25	-25	-25	-20
-20	-20	-20	-20	-25	-30	-30	-25	-20	-20	-15	-15	-15	-15	-15	-20	-20
-20	-20	-10	-5	-5	-5	-10	-10	-10	-5	-5	-10	-15	-10	-10	-10	-20
-20	-20	-20	-20	-20	-10	-5	-5	-10	-15	-15	-15	-15	-15	-20	-20	-20
-20	-15	-20	-15	-15	-20	-20	-20	-20	-20	-20	-25	-25	-30	-30	-25
-25	-25	-20	-20	-25	-30	-35	-35	-35	-35	-35	-30	-30	-30	-30	-35	-35
-35	-40	-35	-35	-35	-35	-40	-40	-40	-40	-40	-30	-30	-25	-20	-15
-15	-20	-20	-25	-25	-30	-35	-35	-20	-5	4	4	4	9	4	0	-25
-49	-54	-49	-49	-59	-59	-59	-54	-49	-59	-79	-98	-108	-108	-118	-137	-132
-113	-98	-93	-93	-83	-79	-64	-59	-49	-44	-40	-30	-25	-20	-20	-30	-35
-40	-40	-40	-35	-30	-35	-30	-30	-30	-25	-30	-30	-30	-25	-25	-30	-20
14	68	136	195	253	302	331	346	326	282	195	92	-10	-118	-181	-235	-327
-435	-522	-586	-654	-801	-947	-1015	-1001	-913	-742	-527	-313	-127	39	160	200	253
380	487	521	463	326	185	92	34	0	-15	-20	-25	-20	-20	-20	-20	-20
-25	-25	-30	-35	-35	-35	-30	-30	-30	-30	-30	-40	-40	-44	-44	-40
-40	-40	-40	-40	-40	-49	-49	-49	-54	-54	-59	-59	-59	-59	-64	-69	-64
-69	-69	-69	-74	-74	-74	-79	-79	-83	-88	-88	-83	-83	-88	-88	-88	-93
-93	-103	-98	-103	-108	-108	-118	-118	-118	-118	-122	-137	-137	-142	-142	-137	-147
-152	-157	-157	-157	-166	-166	-176	-176	-181	-186	-196	-196	-201	-205	-220	-230	-235
-240	-240	-249	-254	-254	-259	-264	-279	-283	-288	-293	-298	-303	-308	-313	-313	-313
-318	-322	-322	-322	-322	-327	-322	-313	-313	-298	-293	-279	-269	-254	-244	-235	-220
-201	-186	-176	-166	-157	-157	-142	-122	-118	-108	-98	-88	-79	-79	-79	-69	-64
-59	-59	-59	-49	-44	-44	-49	-49	-49	-49	-40	-40	-40	-35	-40	-40	-40
-49	-49	-44	-40	-44	-49	-44	-40	-44	-54	-59	-59	-59	-59	-64	-64
-59	-59	-64	-64	-74	-74	-69	-69	-74	-69	-64	-69	-69	-69	-74	-74	-74
-79	-79	-64	-59	-59	-69	-69	-69	-74	-69	-69	-64	-59	-54	-59	-69	-64
-64	-69	-69	-69	-64	-59	-54	-59	-64	-64	-59	-59	-59	-59	-59	-54	-49
-49	-59	-59	-59	-59	-49	-59	-59	-54	-49	-44	-49	-59	-59	-59	-54	-59
-59	-59	-59	-49	-54	-59	-64	-64	-64	-64	-64	-64	-64	-64	-64	-64	-69
-69	-69	-69	-69	-64	-64	-64	-74	-74	-74	-79	-79	-74	-79	-74	-74	-74
-79	-79	-79	-79	-79	-79	-79	-79	-74	-83	-83	-83	-79	-74	-74	-79
-74	-69	-64	-64	-59	-64	-74	-74	-69	-69	-74	-74	-64	-59	-49	-40	-30

CHAPTER 6

Computer analysis of locomotion

MICHAEL R.PIERRYNOWSKI

1. INTRODUCTION

Human motion has been analysed since the time of Aristotle but only recently have quantitative and objective measures been available to study the biomechanics of locomotion. Locomotion is a complex activity and its analysis can give (i) clinical information relevant to the diagnosis and treatment of a patient's musculoskeletal and/or neuronal disorders and (ii) coaching insights into the improvement of an athlete's performance.

Essentially, locomotion laboratories identify the location of a subject on a body movement functionality scale at a particular time. This scale ranges from complete motor disability (death) to elite athletes with coma, amputation, paralysis, normal and athletic in between. The clinical or coaching teams then attempt to shift the patient or athlete along this continuum.

To determine the precise amount of body movement functionality a particular subject has, a locomotion laboratory data collection and analysis system must meet seven criteria before it can receive wide acceptance.

(i) The testing environment should be as natural and comfortable as possible for the subject to minimize the artificial aspects of the laboratory setting on the locomotor pattern. This implies that the locomotion data should be from free-ranging and unencumbered subjects.
(ii) The system acquires, analyses and archives six data sets (documentation, segment anthropometry, foot-ground contacts, marker kinematics, external forces and point of contact, and electromyography) obtained from each subject.
(iii) The equipment collects data that is both valid and has high precision such that one has confidence in the results obtained. In addition, the system must provide insight into locomotor processes not detectable using visual observation (1).
(iv) Full three-dimensional information is obtained (three translations and three rotations for each segment) in a reference system that is easy to understand (i.e. segment fixed).
(v) The data is not obtained from a single step or stride but is an average (ensemble) over many cyclical patterns. The mean and variance of these data are then compared to representative population means to detect abnormalities, or changes due to imposed interventions.
(vi) The system should be highly automated. This will reduce human subjectivity and in some cases drudgery. The ideal system will also have rapid processing

to provide real-time feedback to the subject and to allow several runs to be performed per day.

(vii) Cost should be low so locomotion analyses can be performed routinely.

Fortunately, in the past 20 years advances in electronics, and in particular computer hardware and software, have been instrumental in the development of automated gait laboratories that meet the above seven criteria. In this chapter the computer hardware and software, and associated gait laboratory equipment used within the Biomechanics Laboratory at the University of Toronto will be discussed. This particular laboratory examines the human body from clinical (patient gait), sport (elite athlete performance), and ergonomic (subject—workplace interaction) perspectives. Subject tailored anthropometry, three-dimensional movement and external force recordings, and acquisition of surface electromyographs (EMGs) are integrated in a software package that processes these input data to provide temporal, kinematic, energy, momentum, and force outputs, both in numerical and graphical form. The integration of data from several sources, and their processing are the essential components of a locomotion laboratory. Computers are required to perform these analyses.

2. COMPUTER SYSTEM AND LOCOMOTION LABORATORY EQUIPMENT

The present locomotion data acquisition system uses an IBM AT microcomputer, consisting of a 8 MHz 80286 CPU, 80287 numeric coprocessor, and 1024-Kbyte RAM. A 1.2 Mbyte floppy disc and a 35 Mbyte hard disc are used for archival and semi-permanent data storage. This microcomputer is attached to several input and output devices, specifically selected for use in a locomotion laboratory. These are discussed below and are diagrammed in *Figure 1*.

2.1 Input devices

Input devices provide a means of entering locomotion data into the computer's memory for subsequent processing and storage. Some of these devices are standard computer peripherals. Others are highly specialized.

2.1.1 *Keyboard*

The keyboard provides a means of inputting textual data into the computer via a standard interface.

2.1.2 *Mouse*

A PC mouse, Bus Plus (Mouse Systems Corporation) is a device that allows the user to quickly point to various items (text within menu selections, critical points on graphs, features on images) under software control.

2.1.3 *A/D converter*

This allows the user to convert continuous analogue (voltage) data to digital form. The IBM AT is attached to a 16 channel, single-ended, ± 10 V, 12 bit (4096 levels), A/D converter (WATSCOPE, Northern Digital, Inc.). Input of foot-ground contact,

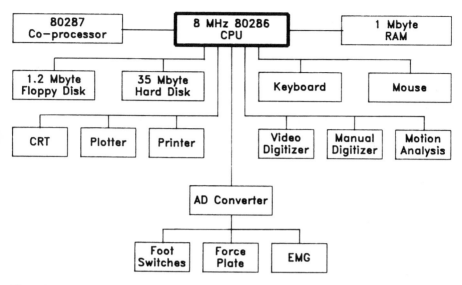

Figure 1. Configuration of the microcomputer and associated input and output devices in the locomotion laboratory.

external force and point of contact, and electromyographic information, provided in voltage form by foot-switches, a force plate, and bioamplifiers, respectively, are channelled through this device. These data are acquired at software selectable rates (maximum of 32 kHz) but are generally obtained at 30−500 Hz.

(i) *Foot-ground contact.* Pressure sensitive switches are devices which open or close an electrical circuit when pressure is applied. Changes in voltage through switches attached to the heel, ball and toe of the foot, which are sensed by an A/D converter, provide a means of recording the time of contact of these respective foot parts with the ground. Foot switches have a variety of configurations and are generally manufactured in-house. A short technical note by Ross and Ashman (2) describes the manufacture of an acceptable thin foot switch to be used in locomotion studies.

(ii) *Force plate.* A Kistler force plate (type Z4852B11), size 0.6 by 0.9 m^2, was mounted flush in the concrete laboratory floor such that an unobstructed 5 m approach and follow through was available. The eight signals evoked by this piezoelectric force transducer are fed under the floor via a 5 m cable to separate charge amplifiers (Kistler 5007Y15). These eight signals are then summed in two Kistler type 5217 summation amplifiers, resulting in signals proportional to ground reaction forces applied to the floor in the forward−back, up−down, and right−left directions. A Kistler dividing amplifier (type 5215Y0117) is then used to calculate the point of force application of the resultant ground reaction force and the free moment about a vertical axis.

(iii) *Electromyography.* Two 4 channel MBS-484 (Moroz Biomeasurement Systems) provide eight bio-amplification units for the measurement of muscle electromyographic signals (raw EMG). Each channel has a 7 position gain control (1,2 and 5 hundred, 1,2,5 and 10 thousand), has a flat frequency response from 8 or 20 Hz (3rd order response) to 1500 Hz (2nd order response), and has a common mode rejection of 95 dB. An advantage of this system is that it employs small pre-amplification modules at the electrode site. This arrangement minimizes signal artefacts due to transmitting low voltage EMG signals through long connecting wires.

Additional analogue processing of the raw EMG signal is possible using a custom built processor which full-wave rectifies and low-pass filters (6 Hz cut off) the signal. This processing is necessary if sampling (A/D conversion) of the EMG signals are desired to be time synchronized to the foot-switch, force plate and motion analysis inputs.

2.1.4. Digitizers

In a locomotion laboratory the two-dimensional (2D) coordinates of selected body endpoints or anatomical features, which are captured on film as they change their location during the locomotor cycle, must be input into the computer's memory. Three devices are used to acquire these data.

(i) *Video digitizer.* A Panasonic WV-1850 high resolution (800 lines) black and white closed circuit television camera provides a EIA standard RS-170 sync image to an OCULUS-200 video digitizer board (Coreco Inc.). This board grabs a 512 × 512 pixel × 128 grey scale representation of the image which is displayed on a Sony PVM-122 video monitor. This device is used to acquire subject tailored anthropometry using a photogrammetric procedure.

(ii) *Motion analysis.* The WATerloo Spatial Motion Analysis and Recording Technique [WATSMART] (Northern Digital Inc.) is a digitizing and motion analysis system which provides the three-dimensional (3D) displacements of up to 64 markers attached to a subject at rates of 19–400 Hz. Using active IR emitting diodes (IREDS) as markers, and four IR sensitive cameras, a direct linear transformation technique is used to reconstruct the two-dimensional camera information into 3D coordinates (3).

(iii) *Manual digitizer.* Input of 2D coordinate data is commonly performed using a digitizer tablet. One such device is the Hewlett Packard 9874A digitizer system which with back projection, high resolution (0.025 mm), and adjustable working surface provide the means to acquire excellent 2D data. A useful feature of the HP9874A digitizer system is the ability of a researcher to digitize in continuous mode. Rather than taking one point at a time, as in recording the position of a marker on the body, it is possible to digitize continuously, so that as the stylus moves across the surface, a steady stream of positions are digitized every 0.1 mm. By outlining the subject, a fair representation of the subjects silhouette can be stored in memory and subsequently analysed or plotted. This enables the researcher to produce an actual reproduction of the subject. This device is used to manually acquire the same data sets automatically captured by the video digitizer and motion analysis systems.

2.2 Output devices

Output devices provide the means to visually display the acquired and processed data in a form convenient, and hopefully aesthetically pleasing, to the end-user.

2.2.1 Graphics

The enhanced graphics adapter (EGA) and monitor (640 by 350 pixels, 16 colours, 2 image pages) was selected as a compromise between sufficient resolution and cost. Ideally an increased pixel density and colour range would be available for the visual presentation of the locomotion data.

2.2.2 Plotters

A Hewlett Packard HP7475A 6-pen plotter produces high quality permanent records, of the locomotion output. Different colours, lettering sizes, symbols and dashed-line fonts are available. In addition, the HP7475A claims high resolution (0.025 mm) and repeatability (0.1 mm) to produce attractive plots.

2.2.3 Printers

A Roland PR-1111A dot matrix printer and a Hewlett Packard Laserjet Series II printer provide draft and high quality fast textual output, respectively.

3. LOCOMOTION DATA ANALYSES

A locomotion laboratory requires six input data sets.
(i)   Documentation.
(ii) Segment anthropometry.
(iii) Foot-ground contacts.
(iv) Marker kinematics.
(v)   External forces and point of contact.
(vi) Electromyography.

With these, a complete biomechanical analysis can be performed. The latter four data sets must be synchronously collected, at a data rate sufficiently high to capture the subtle nuances of the movement being studied. This data rate is generally 30−50 Hz for gait, to 200−500 Hz for fast activities such as sprinting and jumping. These data are then processed to generate seven output data sets.
(i)   Temporal analyses.
(ii) Segment and joint kinematics.
(iii) Segment energies.
(iv) Segment momenta.
(v)   Joint moments and forces.
(vi) Joint and muscle powers.
(vii) Muscle forces.

It is not the purpose of this chapter to define the precise procedures used to measure, analyse and explain locomotor movements. It will however, outline the flow of input and output data present within the locomotion laboratory (see *Figure 2*) and refer the reader to appropriate literature. Further specific details can be gleaned from the books by Winter (4) and Dainty and Norman (5).

At this point a comment about reference systems is in order. The input and output data sets acquired and generated by a locomotion laboratory should ideally adhere to

Computer analysis of locomotion

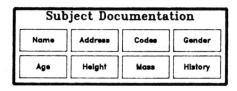
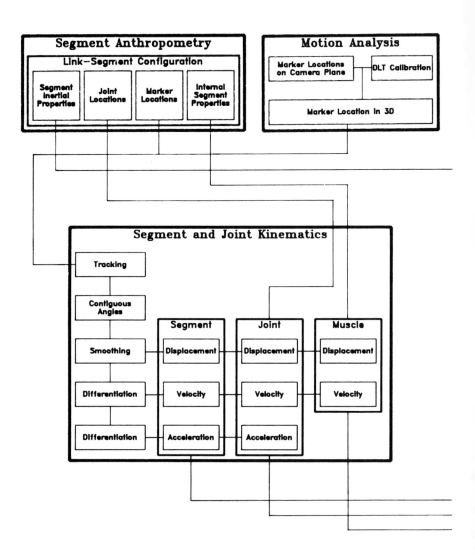

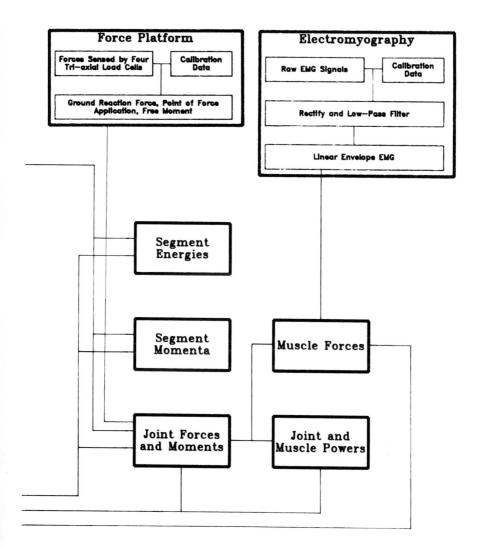

Figure 2. Relationships between the 6 input data sets (documentation, segment anthropometry, foot−ground contact, marker kinematics, external forces and points of contact, electromyography) and 7 output data sets (temporal analyses, segment and joint kinematics, segment energies, segment momenta, joint moments and forces, joint and muscle powers, muscle forces) available from a locomotion session.

an accepted frame of reference (FR). Unfortunately, at least two FRs are in common usage. The first defines the X, Y and Z directions to be forward, up and to the right, respectively, whilst the second defines forward, to the left, and down to be the X, Y and Z directions. Even more confusing is the method by which researchers report rigid body rotations since 12 rotation systems exist (space versus body fixed times the 6 possible rotation orders since rotations are not communitive). In our particular laboratory the following convention is used. Each body segment is positioned relative to a space fixed reference system such that forward ($+Xt$), upward ($+Yt$), and to the right ($+Zt$) translations, and body-fixed flexion ($+Zr$), abduction ($+Xr$), and internal rotation ($+Yr$) rotations of the right side of the body, in that order, from the segment originally oriented in the vertical anatomical position are used (6).

3.1 Input data

3.1.1 Documentation

Knowledge of some pertinent subject data is generally required for documentation and database indexing purposes. Information regarding name, address, subject code(s), height, mass, gender, birthdate and place, tend to be generic. For patients a brief medical history may be included, whilst for athletes training and competition results are recorded. These data are used for cross-referencing purposes such that sample means and variances (i.e. all patients and athletes, male, 20−30 years, with right anterior cruciate ligament injury) can be extracted from the computer database.

3.1.2 Segment anthropometry

Knowledge of the shape, size, configuration and composition of the subject in a standard reference position (SRS) is required to perform locomotion analyses. The segment anthropometry files store this information in five sections.

(i) *Link-segment configuration.* The configuration of the link segment model which specifies which parts of the body are to be considered and their connections. The ideal system would allow the user to define any desired model of the subject. For locomotion studies focusing on the lower extremities a seven link-segment model is defined: head-arms-trunk, two thighs, two legs, two feet, connected by two hips, two knees and two ankles. Further model refinements could include separate forefoot and rearfoot segments.

(ii) *Segment inertial properties.* Several mathematical techniques have been proposed for determining the inertial properties (mass, mass centroid, moment of inertia) of living segments. In this laboratory a photogrammetric technique similar to that developed by Jensen (7) uses two orthogonal views of each segment. Each modelled segment's endpoints and outline are manually or video digitized, rotated about the proximal endpoint to a vertical position, then mathematically sectioned into 50 zones of equal thickness. Using these outline points the radii for each zone are determined and standard formulae used to calculate zone volumes and volume centroids. Using archival segment density values (8) the segment's inertial properties are calculated. These data, calculated when the segment is in a vertical orientation, are then manipulated such that the subject in the SRS.

(iii) *Joint locations*. The photogrammetric procedure outlined above gives the 3D location of the joint centres when the subject is in the standard position.

(iv) *Marker locations*. Identical procedures to those used for the joints are used to calculate the 3D location of the attached segment markers, which are used to measure the kinematics of the link-segment model.

(v) *Internal segment modelling*. Locomotion analyses which attempt to predict muscle forces require knowledge of the muscle's line of action which is the physical location of the muscle within the subject. This is generally defined as a set of up to six 3D points fixed (SRS) in proximity to the skeletal members, these data are used to derive the moment arms, lengths, and velocities of the muscles. The paper by Brand *et al.* (9) outlines one such model of the human lower extremity muscular anatomy. Data defining the size, shape and structure (fibre length, tendon length, angle of pinnation, muscle mass, tendon cross-sectional area, fibre composition) of the modelled muscles are also required. These data are generally derived from archival, cadaver based sources (10), but imaging techniques such as CAT or NMR scans can provide some useful living subject data (11).

3.1.3 *Foot-ground contacts*

To average (ensemble) locomotion data across a series of strides a knowledge of discrete instants within the locomotor cycle are required. These are normally obtained from the sequencing of the foot−ground contacts such that the time between successive right heel contacts define the stride durations. Using switches attached to the heel, ball and toe of the foot, also allows discrete instants to be defined within the locomotor cycle to be used for averaging purposes. Since the foot−ground contact data is available as a time series of A/D converted voltages, computer software is written and executed to identify changes in these voltage (numerical) states. The temporal precision of these contacts is therefore plus or minus half the sampling period.

3.1.4 *Marker kinematics*

The motion analysis system provides the 3D movements of the markers attached to the modelled segments as the subject moves in space (GRS). These data, in conjunction with the segment anthropometrics, are used to define the segment and joint kinematics.

3.1.5 *External forces and point of contact*

The force plate and companion electronic unit provides voltages proportional to the forces applied to the ground in the forward−back, up−down, and right−left directions (Fx, Fy, Fz), the point of force application (ax, ay, az), and the free moment about a vertical axis (My'). These signals are A/D converted, time synchronized to the foot−ground contact, segment kinematic, and electromyographic locomotion data and used in the joint moment and force, joint and muscle power, and muscle force analyses.

3.1.6 *Electromyography*

The electromyographic equipment provides raw, rectified, low-pass filtered, or integrated EMG information from the monitored muscles. These data are used in muscle force analyses either to check model results or as an essential model input.

3.2 Output data

A great variety of data may be obtained, either in numerical or graphical form, from a locomotion laboratory using a smorgasbord of data processing techniques. These fall into seven main categories as outlined below.

3.2.1 *Temporal analyses*

The foot−ground contact data provides the times during the locomotion cycle when the heel, ball or toe contact the ground. From these data is is easy to calculate the mean and variance of the stride period, stance−swing times, foot-contact times etcetera. In addition, since the time when successive right heel contacts occur are known this stride period can be time normalized to 100% and 51 equal time periods defined (0 to 100, by 2%). All of the locomotion laboratory output data could then be processed, using data interpolation algorithms such as the spline (12), to calculate values at these time instants for numerical or graphical output. Although this normalization procedure is not required when examining one subject performing one locomotion sequence it is if one desires to compare locomotor changes across trials and/or subjects stored in the computer's database.

3.2.2 *Segment and joint kinematics*

The motion analysis system gives the 3D time varying positions of those markers located on the segments during the anthropometry procedure. Assuming that at least three non-colinear points per segment are available in both the anatomical (SRS) and space (GRS) frames of reference, rigid body transformations (three translations and three rotations) are defined which move the segments from the SRS to the GRS for each time during the locomotion cycle as desired (13). Knowing the location of the segment in the GRS the internal segment modelling values (joint centres, muscle and ligament lines of action, mass centroids) are also defined. The lengths of the muscle and ligaments and their moment arms relative to the joint centres are then calculated.

These transformations give the linear and angular displacements of the segments as a function of time. These data then undergo a two stage massage. First, the angular data is transformed to avoid discontinuities as the angle moves from 359 to 0 degrees or back. These data are forced contiguous by adding or subtracting 360 degrees as the segments move through these critical configurations. Second, noise inherent within the linear and angular data are attenuated using one of several commonly available smoothing methods [polynomial curve fitting, Kalman or Butterworth digital filters, cubic or quintic spines, Fourier decomposition and reconstruction, (3)]. The contiguous and smoothed data is then differentiated twice using 3rd or 5th order finite difference methods, or using calculus on the fitted equations (polynomial, spline, Fourier), to produce the linear and angular velocities and accelerations of the segments. Similar procedures are employed to calculate the velocities and accelerations of the joint centres, changes in the muscle and ligament lines of action, and mass centroids.

3.2.3 Segment energies

The calculation of the potential and the linear and angular kinetic energies of the segments give insight into the economy of human movement when compared with relative metabolic costs (14). Since, this procedure only requires the segment kinematics and inertial properties data, and not the ground reaction force or electromyographic data, this suggests that energy analyses are a simpler, more robust analysis when compared to joint force, muscle force, or power calculations. In addition, these analyses can be extended to examine internal work and segment energy transfers (15). Unfortunately, these analyses are fraught with controversy as discussed in (14).

3.2.4 Segment momenta

The estimation of the linear and angular momenta of the segments of the modelled segments, and the calculation of the local and transfer segment contributions to the total body momentum, constitutes a valuable tool for the understanding of somersaulting—twisting airborne movements (16).

3.2.5 Joint moments and forces

To determine the joint moments and forces during activity, the mass, inertia tensor, joint displacement and mass centroid displacement and acceleration, for each segment, are required. These are outputs of the segment anthropometry and kinematic analyses. In addition, the external forces and point of contact input data are also required. Using a free body diagram of the link segment model, cut at any joint, the equations of motion for the remaining 'segment' may be stated in terms of forces and moments. D'Alembert's principles of motion are then used to equate the X, Y and Z joint forces and moments with the gravitational, inertial and external forces (17). These values can then be resolved into components adhering to the appropriate SRS's such that joint normal and shear forces are available for perusal.

3.2.6 Joint and muscle powers

Having calculated the joint kinematics, forces and moments the passive rate of energy transfer across the joint centre and the generation, absorption and transfer of muscle power can be estimated. The paper by Robertson and Winter (18) should be consulted for appropriate equations and examples.

3.2.7 Muscle forces

Knowing the configuration and properties of the anatomical system, and the joint forces and moments, it is not a trivial task to assign forces to the muscles modelled to cause the movements observed. This is because of the mechanically redundant nature of the musculoskeletal system. There are often more muscles present than are required to produce any displacement pattern evident from motion data, and the classical equations of kinetic analyses do not permit an unique solution of the muscular forces crossing the joints. To circumvent this problem four categories of techniques have been reported in the literature to assign muscle forces.

Computer analysis of locomotion

(i) *Reduction technique.* This technique solves an initially indeterminate problem by reducing the complexity of the system by grouping muscles and simplifying the anatomy. The work of Morrison (19) explains the details of this procedure.

Table 1. Input data.

(i)	*Documentation* 8 items	8 items
(ii)	*Segment anthropometry* (7 segments * 16 inertial properties) + (7 segments * 6 markers * 3D) + (10 muscles * 6 anthropometry items) + (10 muscles * 6 line of action points * 3D) + (6 joints * 3D)	496 items
(iii)	*Foot-ground contacts* 2 feet * 250 frames	500 items
(iv)	*Marker kinematics* 7 segments * 6 markers * 250 frames * 3D	31500 items
(v)	*External forces and point of application* 6 values * 250 frames	1500 items
(vi)	*Electromyography* 10 muscles * 250 frames	2500 items
	Total input data volume	36 504 items

Table 2. Output data.

(i)	*Temporal analyses* 13 items	13 items
(ii)	*Segment, joint and muscle kinematics* 7 segments * 6 * 102 values * 3D + 6 joints * 3 * 102 values * 3D + 10 muscles * 6 * 102 values	24 480 items
(iii)	*Segment energies.* 7 segments * 3 energies * 102 values * 3D	6426 items
(iv)	*Segment momenta* 7 segments * 4 items * 102 values * 3D	8568 items
(v)	*Joint moments and forces* 6 joints * 2 items * 102 values * 3D	3672 items
(vi)	*Segment and joint powers* 7 segments * 102 items * 3D + 6 joints * 102 items * 3D	3978 items
(vii)	*Muscle forces* 10 muscles * 102 items	1020 items
	Total output data volume	48 157 items
	Total number of entries in database	84 661 items

Based on a mean and variance output datum calculated every 2% of the locomotion cycle.

(ii) *Optimization method.* This method distributes the force in such a way as to minimize some objective such as total muscle force, stress, power, energy cost, fatigue or maximize endurance. (See 20,21,22 and 23 for method details.)
(iii) *Physiological models.* These methods use anatomical, physiological, mechanical and neurological data to resolve the indeterminancy problem (24,10).
(iv) *EMG driven models.* These models, as reported by Hof (25) and Onley (26) use the measured muscle activation patterns as additional information to assign force.

4. CONCLUSION

The need for computers in locomotion laboratory data input and analysis is best summarized by giving a typical gait example. Consider a seven segment human model (pelvis, thighs, legs, feet) each with six markers affixed, six joints (hips, knees, ankles), and 10 muscles (right iliopsoas, glutei, tensor fascia latae, hip adductors, vasti, rectus femoris, hamstrings, gastrocnemius, soleus, tibialis anterior). The motion analysis, foot−ground contact, external forces and point of contact, and the electromyographic data is collected at 50 Hz for 5 sec (250 frames). The volume of data acquired and output is as shown in *Tables 1* and *2*.

Hopefully this example illustrated in *Tables 1* and *2* establishes that computers are essential to acquire, process and store the volume of data available from one locomotor trial. If several subjects, performing multiple trials, are to be accessed, so that between subject and movement comparisons can be performed, logical data structures, and efficient database search and retrieval algorithms must be employed. This will be one of the entities inherent within future integrated locomotion laboratory computer systems.

5. REFERENCES

1. Saleh,M. and Murdoch,G. (1985) *J. Bone Joint Surgery,* **67B**, 237.
2. Ross,J.D. and Ashman,R.B. (1987) *J. Biomechanics,* **20**, 733.
3. Walton,J.S. (1981) *Close-range cine-photogrammetry. A generalized technique for quantifying gross human motion.* Dissertation Abstracts International, Number 8120471.
4. Winter,D.A. (1979) *Biomechanics of Human Movement.* John Wiley and Sons, Toronto, Canada.
5. Dainty,D.A. and Norman,R.W. (1987) *Standardizing Biomechanical Testing in Sport.* Human Kinetics Publishers. Inc. Champaign, IL, USA.
6. Tupling,S.T. and Pierrynowski,M.R. (1987) *Computer Methods and Programs in Biomedicine,* **25**, 527.
7. Jensen,J. (1978) *J. Biomechanics,* **11**, 349.
8. Dempster,W.T. (1955) *Space requirements for the seated operator (WADC-TR-55-159).* Aerospace Medical Research Laboratory, Wright-Patterson Air Force Base, OH, USA.
9. Brand,R.A., Crowninshield,R.D., Wittstock,C.E., Pedersen,D.R., Clark,C.R., and van Krieken,F.M. (1982) *J. Biomechanical Engineering,* **104**, 304.
10. Pierrynowski,M.R. and Morrison,J.B. (1985) *Mathematical Biosciences,* **75**, 69.
11. Hudash,G., Albright,J.P., McAuley,E., Martin,R.K. and Fulton,M. (1987) *Medicine and Science in Sports and Exercise,* **17**, 417.
12. Vaughan,C.L. (1980) *An optimization approach to closed loop problems in biomechanics.* Dissertation Abstracts International, Number 8028300.
13. Miller,N.R., Shapiro,R. and McLaughlin,T.M. (1980) *J. Biomechanics,* **13**, 535.
14. Williams,K.R. and Cavanagh,P.R. (1983) *J. Biomechanics,* **16**, 115.
15. Pierrynowski,M.R., Winter,D.A. and Norman,R.W. (1980) *Ergonomics,* **23**, 147.
16. Dapena,J. (1978) *J. Biomechanics,* **11**, 251.
17. Bresler,B. and Frankel,J.P. (1950) *Am. Soc. Mechanical Engineers,* **48-A-62**, 27.
18. Robertson,D.G.E. and Winter,D.A. (1980) *J. Biomechanics,* **13**, 845.
19. Morrison,J.B. (1968) *Biomedical Engineering,* **3**, 164.

20. Seireg,A. and Arvikar,R.J. (1973) *J. Biomechanics*, **6**, 313.
21. Crowninshield,R.D. (1978) *J. Biomechanical Engineering*, **100**, 88.
22. Pedotti,A, Krishnan,V.V. and Stark,L. (1978) *Mathematical Biosciences*, **38**, 57.
23. Hardt,D. (1978) *J. Biomechanical Engineering*, **100**, 72.
24. Hatze,H. (1976) *Mathematical Biosciences*, **28**, 99.
25. Hof,A.L. (1984) *Human Movement Science*, **3**, 119.
26. Onley,S.J. and Winter,D.A. (1985) *J. Biomechanics*, **18**, 9.

CHAPTER 7

Active physical therapy—computerized rehabilitation

JERROLD S.PETROFSKY

1. INTRODUCTION

In much of the hospital environment, computers have come into common use. For example, in the modern clinical laboratory computers are used not only to calculate results and store data but automatically to calibrate instruments such as blood gas and pulmonary function analysers. Historically, computers have seen intensive use in the Emergency Room and in the Intensive Care monitoring station. The advantage of using computers in the hospital environment is that they can continually monitor vital signs from patients and store data. Of all of the areas of hospital use including hospital administration, record keeping, admissions and surgery as well as those cited above, the last area of the hospital environment to see an influx of computer use has been rehabilitation. Automating the physical therapy environment has made the job easier for the physical therapist, but in addition, the digital computer allows many new types of technologies to be applied in the rehabilitation setting. One of these is active physical therapy involving the use of electrical stimulation. This chapter departs from the normal bench manual approach to emphasize the role of the computer in allowing an old idea to breach technical barriers.

2. FUNCTIONAL ELECTRICAL STIMULATION

There have been a number of attempts to restore spinal cord function in paralysed patients, but these have not been notably successful. Goldsmith et al. (1) described attempts to restore spinal cord injury (SCI) impaired function by replacing lost circulation with metal transplants. Naftchi (2) and Faden et al. (3) have tried to achieve similar results with medication. Perkins et al. (4) and Kao (5) have attempted restoration by nerve tissue transplants. Thus far, such attempts have only been partially successful in selected patients whose conditions are specifically amenable to such treatment. Greater success has been achieved in what has been called rehabilitation engineering, with the application of engineering principles to the design and development of artificial systems that can substitute in part for the lost central command function. This effect has resulted in significant progress in producing useful contraction of paralysed muscles in SCI patients. Movement produced in this way can be variable to the patient.

If the SCI is high enough in the spinal column, alpha motor neurons supplying the paralysed muscle to the lower part of the body are not damaged. Therefore, the affected muscles did not undergo denervation atrophy and loss of muscle mass and strength is due to simple disuse atrophy. They retain the ability to contract in response to electrical stimulation. Such functional electrical stimulation (FES) has been used in the research environment for many years. Tenkoczy et al. (6) and Zealer and Debo (7) have described and discussed the use of functional electrical stimulation in detail.

A number of techniques have been used in experimental functional electrical stimulation. These have included the use of intramuscular wire electrodes (8,9), sleeve electrodes (10−12) and surface electrodes (13,14). Scott (14) pointed out that the surface electrodes require high current densities and that this could cause skin burns.

Reswick and Vodovnik (15) brought attention to the application of engineering principles and locomotor disabilities in the same group. Crochetiere et al. (16) studied stimulation of skeletal muscle in special situations. Lieberson et al. (17) used FES with open loop control for the first time in locomotor studies to allow a paralysed individual to stand and Petrofsky and Phillips (18) extended the use in producing closed loop controlled leg movement in paralysed persons. The latter closed loop control (19,20) allowed the restoration of smooth coordinated movement to muscles with surface electrical stimulation.

There are a number of studies indicating that FES-induced exercise can directly affect atrophied paralysed muscle. It has been found that stimulation in contraction of skeletal muscle increases the speed of contraction and oxidative metabolism (21−23). Continued FES-induced exercise in muscles restored the normal distribution of fast and slow twitch muscle fibres (24−27). Cooper et al. (28) implanted sciatic nerve stimulators, and found that after only ten weeks of stimulation weak and paralysed muscles increased in strength up to 25%.

Unfortunately, these studies have varied with respect to stimulus parameters, patient condition and conditions of the study. The studies were also done without control of the motion produced. In most cases, these studies neglected to measure the effects of electrical stimulation on the cardiovascular system as well as the respiratory system. It has only been in recent years that more studies have been attempted.

The relationship between exercise and body fitness is well known. Alam and Smirk (29) observed a blood pressure increase with voluntary isometric skeletal muscle contraction. These observations have been extended to the effects of lack of exercise on the body condition of paralysed persons (30−33).

McCloskey and Mitchell (34) worked with electrical stimulation of cat muscles and observed an increase in arterial pressure with such stimulation. Lind et al. (35) and Petrofsky et al. (36) made this same observation in humans and animals, and an increase in arterial pressure with electrical stimulation of paralysed muscles of the quadriplegic patient has been reported.

There are studies which indicate that SCI paralysed patients, especially those with impaired autonomic function, are deprived of some of the general health benefits of physical activity (*Figure 1*). These health benefits would include reversal of the severe osteoporosis associated with paralysed muscle (37) which itself causes a number of medical problems. As many as 10% of SCIs experience fractures. Benassy (38) observed that bone healing was not a problem in paralytics, which implies that the primary difficulty is one of disuse osteoporosis *per se*.

There have been numerous studies indicating that the forced inactivity of the paralytics results in a reduction of cardiovascular function. This work has been reviewed by Davis

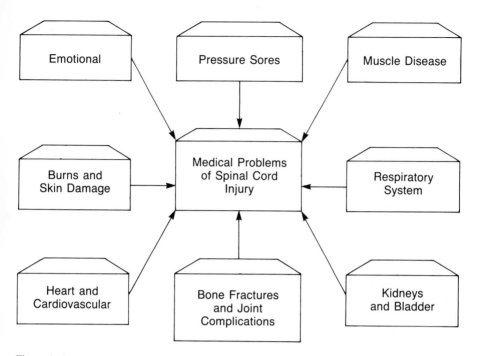

Figure 1. Causes of spinal cord injury.

et al. (39). Nilsson *et al.* (40) described general cardiovascular deconditioning in paralysed persons. Hjkeltnes (41) observed a large difference in AV oxygen in lower limb paralytics, and considered this an indication of impaired circulation associated with lower body blood pooling. Clausen (42) in reviewing the cardiovascular effects of exercise noted that the literature supported the concept of reduced circulatory function in paralytic patients. Shephard (43) pointed out that the normal methods of assessment of physical fitness cannot be used in paraplegics. Davis *et al.* (39) cited a number of researchers who have studied the use of armchair ergometers for upper body exercise to enhance fitness in paralysed persons. Increases in fitness estimated at 20—30% have been observed in exercising humans with upper body exercisers. Knutson *et al.* (44) observed an increase in work capacity and a decrease in heart rate response in paralytics when they carried out an exercise program. Coutts *et al.* (45) in well-designed experiments studied cardiopulmonary function and oxygen uptake in paraplegics and quadriplegics using wheel chair ergometry, and noted a distinct improvement with exercise. Huang *et al.* (46) compared paraplegics with normal, able-bodied individuals, and reported significant differences in fitness. All of these researchers used wheel chair ergometry as the exercise medium.

Bergofsky (47) observed the diminished thoracic compliance in patients with high spinal cord injuries and hypothesized that they may be due to a loss of intercostal function. Axen (48) studied the responses of paralytic patients to imposed pressure loading of the lungs, and did not observe significant differences between normal subjects and paraplegics.

In addition to the dysfunction of the autonomic nervous system associated with paraplegia and quadriplegia, there is a dysfunction of thermal regulation as well. Claus-Walker and Halstead (49) have extensively reviewed the literature on partial decentralization of the autonomic nervous system including thermal regulatory effects. Downey et al. (50) showed that shivering occurred in quadriplegics when the core temperature was lowered, indicating that the deep thermal receptors or certain thermal receptors are still operable. Stimulation of the skin temperature receptors by cold increased the temperature at which shivering occurred in paralytics and, in 1973, they extended these studies to show that central cooling increases hand blood flow in paralysed persons. There has been little work in assessment of thermal regulatory function in paralysed patients. In our own work (51) it has been shown that people who are paralysed from spinal cord injury, both paraplegics and quadriplegics, had reduced tolerance to work and heat. The workload in this case was arm crank ergometry at an intensity of 25 Watts. Quadriplegics could not exceed 15 min of work at environmental temperatures above 50°C with a 50% relative humidity without going into heat syncopy.

As seen from the above studies, a number of investigators have examined the use of electrical stimulation as a means of restoring strength in skeletal muscle. In at least the last 30 years, electrical stimulation has been applied by the use of two electrodes above the muscle, and stimulating it with different patterns of electrical current. These studies have ranged from altering the amplitude of stimulation to altering the frequency and pulse width. Muscles have been stimulated to tetanic frequencies in contrast to turning the muscle off and on for various periods of time during the day (8,25,52,26). The results of these experiments have been interesting but not very encouraging. Generally, they seem to show that although some strength can be restored to paralysed muscle by electrical stimulation, the increase is small compared to that which can be obtained with voluntary physical training. Although endurance also increases, even with months to years of electrically induced exercise such as that described above muscles still fatigue very rapidly. Fatigue lasts for long periods of time. This is in direct contrast to voluntary activity where muscles fatigue rapidly but then recover rapidly as well (53).

The common modality with all of these types of training is that muscles are stimulated in a very non-physiological manner. One of the primary uses of computer technology in the last decade has been to stimulate muscle in a more physiological manner to allow smooth coordinated movement for exercise.

It has been well known in the exercise literature (53,54) that training of skeletal muscle is specific for the kind of exercise being accomplished. For example, simply lifting a weight up and down a few times a day does not increase muscular strength. Increasing muscle strength requires that progressive resistance exercise be accomplished. To develop strength, a weightlifter must lift weights equivalent to approximately two thirds of his maximum strength at very slow rates of lifting up and down. This type of exercise (isokinetics) optimizes the strength conditioning of skeletal muscle. Mueller (55) demonstrated that if weights were lifted at approximately half to two thirds of the muscle's maximum strength very slowly up and down, muscles would develop strength rapidly. In contrast, activities during which the muscle works rapidly but at low loads such as running are associated with only small increases in muscle strength (53). Therefore the type of metabolic load induced in the muscle can be instrumental in determining the type of changes that occur biochemically in skeletal muscle with long-term physical training. It has been in the last ten years, then, that computer controlled exercise involving weight lifting and aerobic exercise such as bicycling have finally enabled paralysed individuals to begin to exercise in a more natural manner and, as described below, achieve some of the beneficial results of exercise not seen before.

In addition to exercise, computer controlled electrical stimulation can be used in the rehabilitation setting for walking as well. In 1961 Lieberson stimulated the quadriceps muscle of a paralysed person, causing the person to stand from a sitting position (17). Kralj and his colleagues have been working since the 1970s using stimulation of reflex loops to initiate walking (56–59). In their work, standing was achieved by stimulating the quadriceps muscles. Since the hip muscles were not stimulated, a C posture was evident due to the extension of the back for balance. Walking was initiated by eliciting a flexion–withdrawal reflex, coupled with pulling the body forward by arm motion, allowing a rudimentary walking movement to occur. Thoma and his colleagues (60) used a multichannel electrode placed around the sciatic nerve in the leg to allow the paralysed person to flex or relax the leg. Using four point gait, rudimentary walking was achieved by alternately swinging the legs through while stimulating the opposite leg, similar to Brindley's electrical splinting of the knee (61).

Embedded electrodes have also been used by Mortimer, Peckham and Marsolais at Case Western Reserve University in Cleveland and the Cleveland V.A. Hospital. This group used fine wire electrodes which were inserted percutaneously into the muscle. They were used in hand control systems to restore limited hand function to quadriplegics (8) and for multichannel stimulation of a variety of muscle groups in the lower part of the body for computer-controlled walking (62).

All of these approaches are similar in that they use open loop control. In such control, the muscles are stimulated in a pre-set pattern with no feedback of information from the joints as to the position of the limbs or pressures generated by the muscles. If the leg is stopped by dragging the foot on the floor or by hitting some obstruction, the sequence continues without interruption. Closed loop control systems allow for adjustment to actual motion through feedback of muscle tension or length to the stimulator (*Figure 2*).

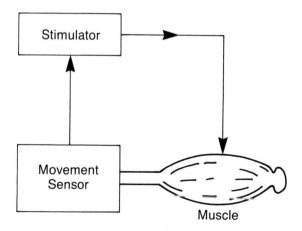

Figure 2. Diagrammatic representation of closed loop control of movement.

Closed loop control systems have been used to control tension during isometric contraction of skeletal muscle (20,63 and 64). This concept has been expanded to include closed loop control of tension, using a digital computer to control a stimulator. Also, closed loop control of the velocity of movement in animal muscle and later in human subjects has been done during exercising and walking. In closed loop control systems, sensors are placed on various parts of the body (e.g. hips, knees and ankles, and pressure sensors on the bottoms of the feet). Information from these sensors is input into the computer by analogue-to-digital converters, and this information is used to determine the timing and amount of stimulation applied to the muscles. As can be seen from the above survey of literature, then, in the last decade or so a wide variety of exercise applications and practical applications such as walking have developed with computer-controlled technology. Further, applications such as computerized biofeedback have also developed to allow people who have incomplete neurological injuries, through computer training, to be able to use their paralysed limbs more successfully. This chapter aims to develop some of these concepts in more detail using spinal cord injury and spina bifida as examples of how computer technology can be developed and used for exercise, feedback, and purposeful movement.

3. EXERCISE DEVICES INVOLVING COMPUTER-CONTROLLED MOVEMENT

As stated in the introduction, the most significant problems associated with neurological diseases such as spinal cord injury, spina bifida or any other injury which results in total paralysis are the secondary medical problems which result from these disorders. Secondary medical problems such as pressure sores, kidney and bladder infections, and spontaneous pathological bone fractures cause a terrible toll in terms of patient costs. In the United States, for example, the average settlement in the courts for medical costs for paraplegia and quadriplegia is over four million dollars, with significantly higher figures for a quadriplegic than a paraplegic. These medical care costs, then, as well as the health care problems associated with paralysis are a tremendous challenge in terms of the application of computer technology to this area of rehabilitation.

In terms of exercise, it is conventionally known that a program involving a combination of anaerobic and aerobic exercise can be quite beneficial in the non-disabled population in terms of improving health and reducing medical costs (*Figure 3*). Such a program typically involves a mixture of two types of exercise. One type of exercise, called anaerobic exercise, involves weight lifting while the other type of exercise, termed aerobic exercise, involves such activities as bicycling, jogging, aerobic dancing and so on. These types of exercise provide a balanced program of muscle building and cardiovascular conditioning. Weight training has been shown to be a superb means of building muscular strength but does little to build the cardiorespiratory axis. In contrast, aerobic exercise such as bicycling does little to build strength in muscle but, because of the high blood flows through muscle and high oxygen consumptions associated with the exercise, build the heart and respiratory system. Therefore, for conventional exercise programs, a combination of the two exercise types is commonly used.

For disabled people, both anaerobic and aerobic exercise can be done in the non-paralysed part of the body. However, in contrast, even with intensive aerobic training

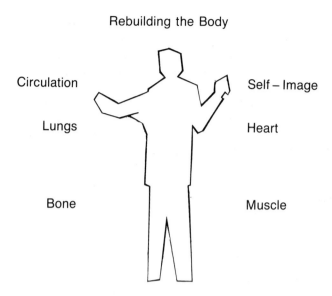

Figure 3. Factors involved in conditioning the body.

in the non-paralysed part of the body, medical problems associated with paralysis are still found. For example, in wheel chair athletes where intensive workouts including wheelchair marathons are done with the upper part of the body, the incidence of pressure sores and kidney and bladder infections is just as high as in wheelchair bound individuals who do not exercise this intensively. Therefore, over the last hundred years or so, electrical stimulation has been used as a therapeutic modality to try to exercise the paralysed parts of the body. Typically, two electrodes are placed on a muscle and stimulation is applied at frequencies which range between 30 and 200 cycles per second. Typical stimulation involves the application of either monophasic or biphasic electrical stimulation with square wave impulses whose amplitude is as high as 300 mA. The most common pulse width of the stimulation is 300 μm pulses. The results of this type of electrical stimulation in the past have been quite poor. When such stimulation has been applied to paralysed muscle, the muscle certainly has contracted. The only exception to allowing muscle to contract has been with low level injuries. With spina bifida in the lumbar area of the spinal cord or low level spinal cord injuries as an example, the cell bodies of the motor nerves which give rise to the axons enervating the skeletal muscle in the lower part of the body are damaged. Once damaged, the axons degenerate by Wallerian degeneration and the result is denervation atrophy of muscle. Denervation atrophy, in most cases, has resulted in an inability to receive electrical stimulation and contract functionally. For these reasons, then, electrical stimulation has only been applied to injuries above the lumbar area of the spinal cord, or in cases of diseases which affect areas above the lumbar area of the spinal cord.

In spite of the fact that muscles do contract, the results of strength training in muscle has been poor. For example, as cited under the survey of literature above, continuous stimulation of muscle results in only a marginal increase in muscle strength even when

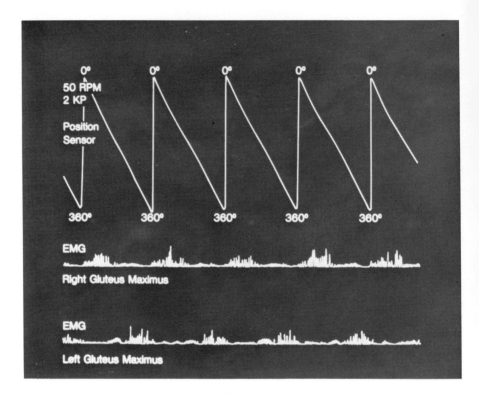

Figure 4. The electromyogram recorded from a non-paralysed volunteer during bicycle exercise. These data are used to model movement for computer controlled electrical stimulation.

stimulation is accomplished for several hours per day on a seven day per week basis. Unfortunately, however, most stimulation has been applied in a non-physiological manner in the past. For example, if a weight lifter were to lift a pencil five days per week for an hour a day, he in fact would lose muscle strength. In contrast, if a weight lifter were to lift heavy weights very slowly up and down (progressive resistance exercise) then muscle strength would increase rapidly. The difference between the two examples cited above is that in the former, no load is applied to the muscle whereas in the latter case a heavy load is applied to a slowly moving muscle. Genetically, muscle responds to load and to the condition that it is trained for. In exercise physiology, it is a common axiom that you train for the type of exercise that you do. Cross-training, that is the ability to train the arm muscles and produce a training effect on the leg muscles, has never been clearly demonstrated. Further, training a muscle for weight lifting trains the muscle to be able to move heavy loads and not to perform endurance functions (53). It is not surprising, with the specificity of training the skeletal muscle to the type of exercise being done, that these earlier experiments failed to produce any significant changes in muscle strength or muscle endurance.

In contrast, when electrical stimulation is used with feedback control, muscle strength builds rapidly. When muscle is stimulated to lift a very heavy load, significant gains in muscle strength can result. An early generation exercise device is the quadriceps

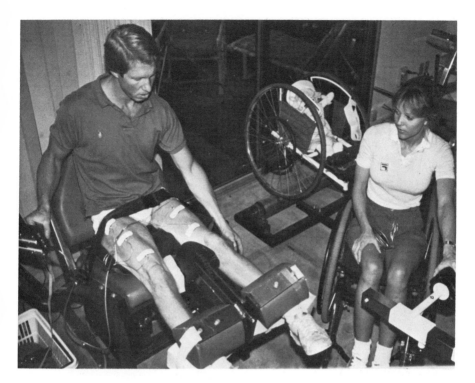

Figure 5. Quadriceps leg extension machine based around a computer controlled weight trainer (Paramount Inc.) modified for SCI use with a computer controlled stimulator and computer interface.

muscle trainer. The weight to be lifted is connected through a series of pulleys to the ankle. When electrical stimulation is applied to the quadriceps muscle, the leg extends. This extension is retarded by the weight which is connected through the pulleys to the ankle. A sensor on the knee provides feedback to the computer as to the position of the muscle. This positional information is used as input for the computer for controlling movement. These types of devices can be quite user interactive. By using speech synthesizers, the computer can actually talk to the individual and, by measuring impedance of the electrodes in real time, alert the individual about improper connection of electrodes to the body, drying of electrode paste, or even a loose electrode. Further, feedback from the computer can be provided through a speech synthesizer in order to encourage the individual to work harder. This type of setup has proven quite helpful in working with adults as well as with spinal cord injured and, in a recent breakthrough in the management of spina bifida, in spina bifida children. When doing rehabilitative exercise with children, the feedback can be crucial in terms of making the child feel comfortable with the equipment.

The actual control associated with movement is derived from analysing the electromyographic records of non-paralysed individuals, as shown in *Figure 4*. As is shown in this figure, the electromyogram was recorded from muscles of the leg during bicycling in a non-paralysed individual. These EMG records are modelled on equations describing

201

Active physical therapy—computerized rehabilitation

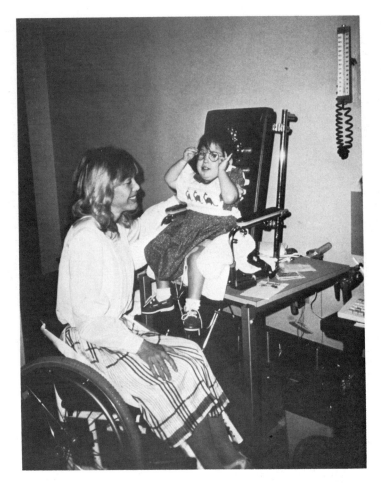

Figure 6. Three-year-old spina bifida child lifting weights with stimulation of the quadriceps muscle at Scottish Rite Childrens Hospital in Atlanta, Georgia in a combined project with University of California, Irvine.

the relationship between the position of the legs as a function of pedal rotation and the onset and intensity of muscle contraction of the major muscle groups of the legs. Once modelled for a given type of activity, closed loop control equations are derived to operate the equipment. One problem with some of the earlier equipment developed in our own laboratory (65–77) was that equipment was too time-consuming to use and did not properly protect the limbs from adverse torque and rotation during movement. For that reason, then, over the last year a new generation of exercise equipment has been developed involving exercising different parts of the body. The analogous situation to this type of equipment would be a health spa, where a person enters and moves from one piece of equipment to another in an exercise circuit. One example of this type of equipment is the quadriceps machine shown in *Figure 5*. Operated by a paraplegic, the quadriceps leg extension machine exercises both legs simultaneously. This type of equipment, when miniaturized, has also proven to work quite well for

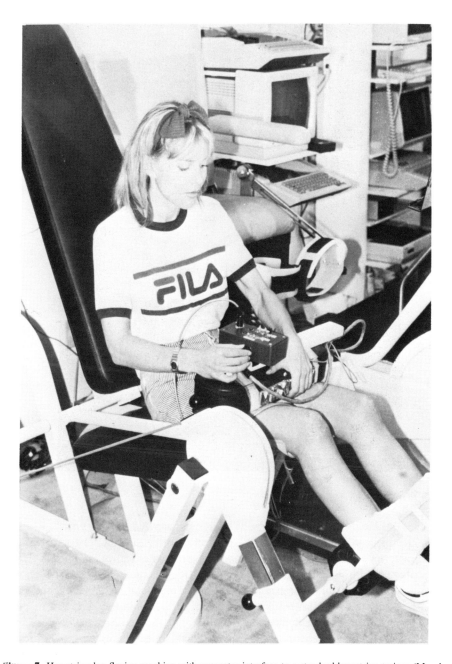

Figure 7. Hamstring leg flexion machine with computer interface to a standard hamstring trainer (Muscle Dynamics Inc.) modified extensively for computer use.

children with spina bifida, as shown in *Figure 6*; computer controlled weight lifting allows them to exercise their paralysed limbs. Spina bifida children whose legs have been paralysed from birth are able for the first time to build up muscle strength in their

Active physical therapy—computerized rehabilitation

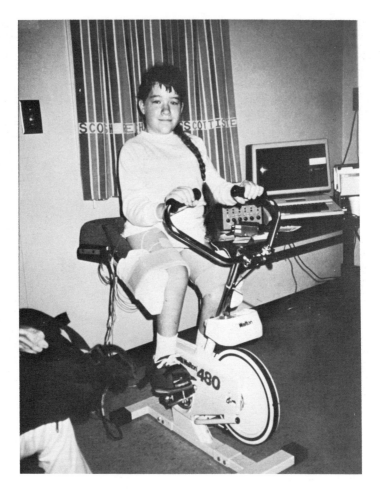

Figure 8. Spina bifida child on exercise bike.

legs and, fitted with a miniaturized version of the computerized walking system, to walk.

Another piece of equipment used in this particular exercise circuit is a gastrocnemius-tibialis anterior exercise machine, and a hamstring trainer as shown in *Figure 7*. These types of weight lifting equipment provide good strength training but, as cited above, provide little in terms of endurance training. Increasing blood flow and mitochondrial density to muscle as well as building the central cardiorespiratory axis is a function of aerobic exercise. Using computer-control algorithms derived from the EMG records as discussed above, an exercise bicycle was developed for aerobic exercise. The bicycle as shown in *Figure 8*, uses special stabilizer bars to keep the knees from rotating in and out during exercise and a high-backed bicycle seat to provide postural support because of the lack of postural muscles in paralysed individuals. These types of devices have been used both for spina bifida children (*Figure 8*) and paraplegics and quadriplegics.

J. S. Petrofsky

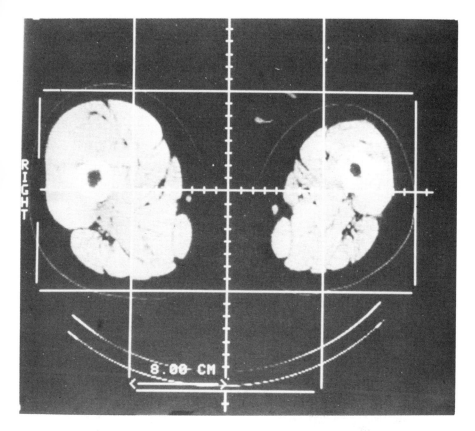

Figure 9. CAT scan of the upper thigh of a paraplegic after three months of electrical stimulation (see description in text).

The weight training devices have been shown to be quite good in terms of building muscle strength. In terms of the actual strength changes, a typical example is shown in *Figure 9*. This figure shows a CAT scan through the thigh. In this particular case, one leg served as a control (non-exercised) while the other leg was exercised for 15 min per day, three days per week over a period of three months. The increase in muscular strength and muscle size is quite evident, as can be seen in this figure. The overall change in physical conditioning, however, far outreaches the simple change in muscle strength. For example, a paraplegic who entered our studies initially, approximately one year later had a dramatic change in appearance. The change in appearance was certainly linked to the exercise itself, but in addition the individual, like many individuals who exercise with this type of equipment, began exercising his upper body voluntarily as well as taking better care of himself. This underscores the psychological benefits that can be derived from computer-controlled exercise. Obviously, this type of exercise does more than simply build muscle strength. For the paralysed individual, exercising the lower part of the body and building up muscle strength once again is crucial in terms of its effect on psychological outlook as well as health care benefits that can be

Active physical therapy—computerized rehabilitation

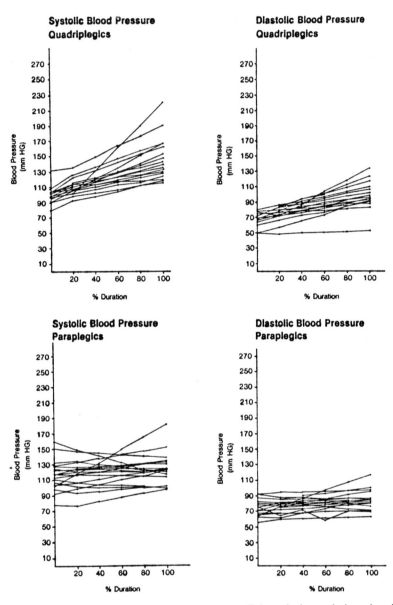

Figure 10. Typical blood pressure responses to computer-controlled exercise in paraplegics and quadriplegics.

derived from the exercise. As might be expected, blood pressure increases during the exercise (*Figure 10*) as does heart rate (*Figure 11*). Cardiac output, particularly on the bicycle, increases rather dramatically, as shown in *Figure 12*. However, there are significant differences in the cardiorespiratory responses of disabled individuals. For example, for the paraplegic, the heart rate increases in the typical manner seen in non-disabled individuals during computer-controlled exercise, that is, there is a linear

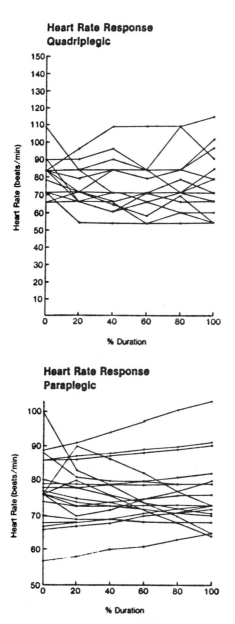

Figure 11. Typical heart rate response to computer-controlled exercise in paraplegics and quadriplegics.

relationship between heart rate and the severity of the workload. In contrast, blood pressure increases marginally if at all during exercise with the lower part of the body, but in the typical manner during exercise of the voluntary muscles of the upper part of the body. In contrast, in a quadriplegic, blood pressure response is normal with exercise in the lower part of the body but abnormal with exercise in the upper part

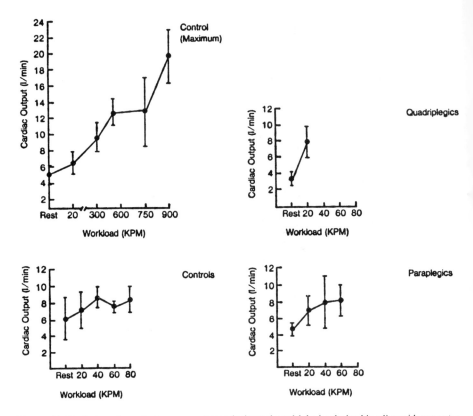

Figure 12. Typical cardiac output response of paraplegics and quadriplegics during bicycling with computer-controlled electrical stimulation.

of the body. Further, heart rate response may be absent in exercise of either the upper or lower part of the body with either computer-controlled or voluntary exercise. The reason for these differences has to do with the level of the injury and its effect on the spinal cord. Since paralysis such as spinal cord injury can affect the autonomic nervous system and sensory pathways as well as the voluntary or alpha motor neuron system of the body, injury levels in the cord become crucial in determining the cardiorespiratory responses to the exercise.

In spite of this variability in response, a great deal of good can be done with computer-controlled exercise in terms of the cardiovascular system, as is shown in *Figure 13* for a group of four quadriplegics (73). The quadriplegics in this group were all users of electric wheelchairs and had very unstable resting blood pressures, as shown on the illustration. During exercise, in the initial phase blood pressure was very high and very unstable. This particular group of individuals had very poor orthostatic tolerance and had a great deal of difficulty in even sitting up for long periods of time. Once the exercise began, however, over a period of approximately four weeks the resting blood pressure increased to normal levels and the exercising blood pressure became stable. This had the overall effect of increasing orthostatic tolerance and allowing these individuals to sit up for extended periods of time without becoming dizzy. This type of autonomic

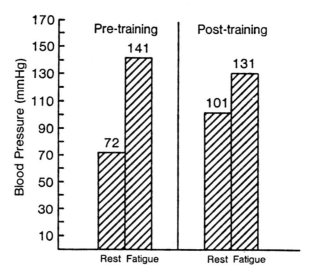

Figure 13. Blood pressure at rest and during exercise before and after a six week period of conditioning on the computer-controlled exercise bike in four quadriplegics.

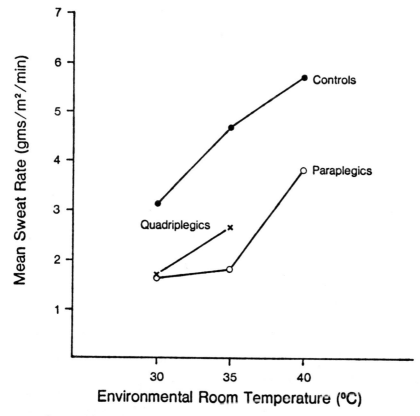

Figure 14. Whole body sweat rates in controls, paraplegics and quadriplegics during 25 Watts of work and exposure to environmental temperatures of 30, 35 and 40°C for 30 min.

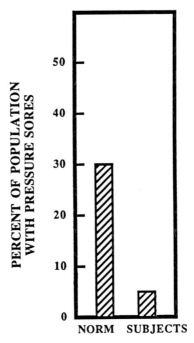

Figure 15. Incidence of pressure sores in people exercising on the computer-controlled exercise equipment and in control subjects.

conditioning is quite good for quadriplegics in terms of improving their lifestyle and sense of well-being as well as their general health.

Because of this damage to the autonomic nervous system, damage to the sweat glands can also result, and inhibit the heat exchange in the body. For example, in *Figure 14* a group of four controls, four paraplegics and four quadriplegics are shown during exercise of the upper body (25 Watts) in heat chambers at 30, 35 and 40°C for 30 min. The results of this exposure dramatically illustrate the differences in sweat rate between the three groups of individuals. With a paralysis of the sweat glands in quadriplegics, heat exchange was severely limited with the overall result of a severe heat exchange problem during exercise at high temperatures.

Obviously, this type of exercise does in fact have health care benefits. There is a dramatic reduction in pressure sores, kidney and bladder infections and bone fractures associated with paralysis for people who exercise with this technology (*Figures 15* and *16*). These figures illustrate the results of 100 paralysed individuals who exercised for a period of one year on the exercise equipment described above. After a period of one year, the health care costs during that year from pressure breakdowns and bone fractures and so on were evaluated and compared with 6000 other people shown as controls who did not exercise on the equipment. The individuals were matched in terms of age and length of time since the injury. The overall result, as shown in *Figure 17*, was a 93% reduction in health care costs for conditions associated with spinal cord injury. Similar studies have not been done for stroke, spina bifida and other types of neurological diseases.

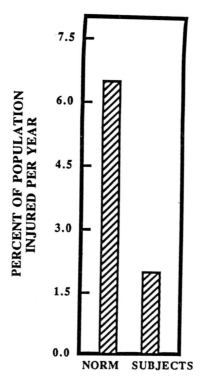

Figure 16. Incidence of bone fractures in 100 spinal cord injured individuals exercising on computer-controlled exercise equipment compared to 6000 control subjects.

4. COMPUTER-CONTROLLED MOVEMENT

The most important use of computer technology has been in the therapeutic setting involving computer-controlled exercise. Computer-controlled movement as described in the introduction has been the subject of a great deal of research in the last few decades. The simplest form of computer-controlled movement is hand control. By using a stimulator above the finger flexors and extensors in a quadriplegic, for example, computer-controlled movement of the hand muscles has been achieved. Such computer control systems have been used with both external electrodes and implantable electrodes, as described in the survey of literature above. The control signal itself is usually derived from the opposite shoulder. A sensor, placed on the shoulder opposite to the hand that is exercising, transduces shoulder movement to finger and wrist flexion and extension.

Computer-controlled walking is a much more complex process. Computer-controlled walking in its simplest form involves the use of biofeedback for retraining combined with electrical stimulation for building up muscular strength. As shown in *Figure 18*, for example, when biofeedback is used to retrain damaged nerve pathways (where there is some conduction of nerve impulses) in combination with electrical stimulation for muscle strength training, significant increases can be seen in the amount of strength controlled voluntarily by muscle (72,78,79). For incomplete injuries or injuries where some nerve activity is still present, biofeedback has always provided a good means

Active physical therapy—computerized rehabilitation

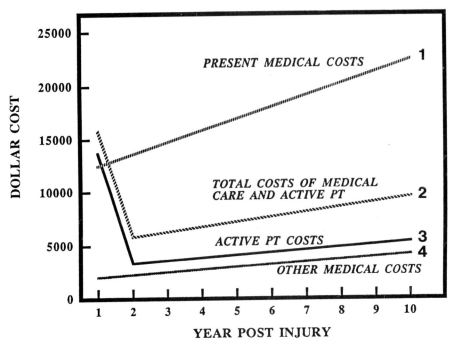

Figure 17. Health care costs associated with being spinal cord injured for the next ten years for people who do not use computer-controlled exercise equipment (1) and individuals who use computer-controlled exercise equipment (2). Graphs 3 and 4 illustrate the health care costs in the individuals who are exercising on the equipment plus the costs of the therapy itself.

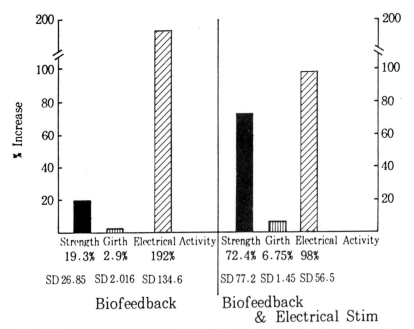

Figure 18. The influence of a combination of biofeedback and electrical stimulation on restoring strength and movement to skeletal muscle in incomplete injuries.

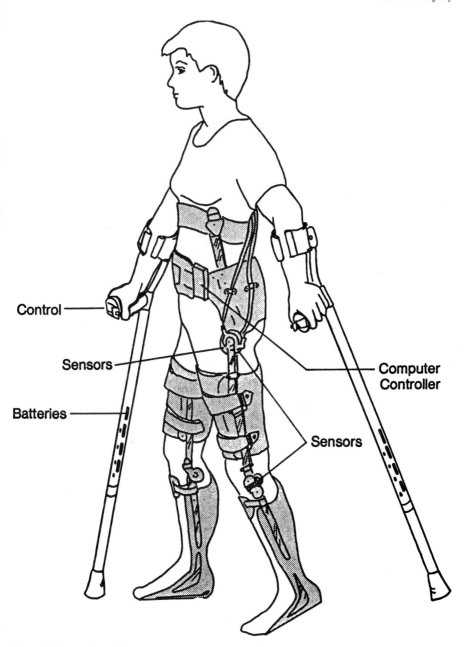

Figure 19. A combined hybrid system of the LSU brace and electrical stimulation.

of retraining damaged nerve pathways and enhancing neural activity. Although paralysis can be seen in many nerve pathways during the acute phase of an injury, over a period of several years muscle control can in fact return to some muscles. However if muscles are too atrophied to move, this control and return of movement is not readily apparent. Therefore, many individuals have certain muscles in their bodies which could in fact

Active physical therapy—computerized rehabilitation

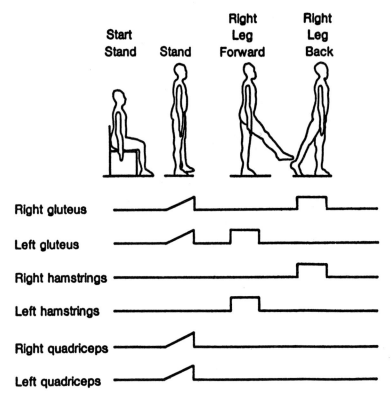

Figure 20. The electrical stimulation control cycle during computer-controlled standing, sitting and walking.

be retrained if their strength could be built up to a significant level. This then is the reason for a combination of biofeedback and electrical stimulation.

Certainly if electrical stimulation for strength training with biofeedback will not allow movement to be returned, full computer-controlled movement can be attempted. There have been a number of systems as described in the survey of literature above which enable paralysed individuals to stand and to take a few steps. In our own research, as an example, we initially tried using a system using computer-controlled electrical stimulation of several dozen muscle groups in the body simultaneously, with closed loop control at the hips, knees and ankles to provide smooth coordinated movement. However, the system was too inefficient because of its reliance upon stimulation of so many muscle groups with so many sensors to provide meaningful movement over long distances. Therefore, electrical stimulation was combined with bracing as a hybrid walking system using the LSU reciprocating gait orthosis in a combined electrical stimulation-bracing system (*Figure 19*). This particular system used computer control during standing, sitting (closed loop) and open loop control during the gait cycle for walking. *Figure 20* shows the control cycle during standing, sitting and walking. The system worked quite well, but suffered from a number of problems. The advantage

J. S. Petrofsky

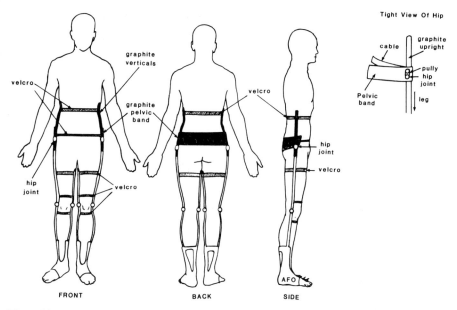

Figure 21. Graphite bracing system used with the hybrid walking system.

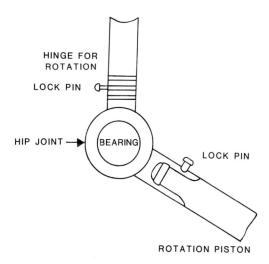

Figure 22. Hip joint design used with the hybrid walking system.

of the brace was that it provided some postural support as well as an emergency backup system to allow individuals to stand if the computer failed. This was a serious problem in earlier walking systems because of medical legal liability. Because of the potential for someone falling and being seriously injured, the bracing system was added to reduce the chance of injury as well as to protect the joints from torque and rotation. With the

Active physical therapy—computerized rehabilitation

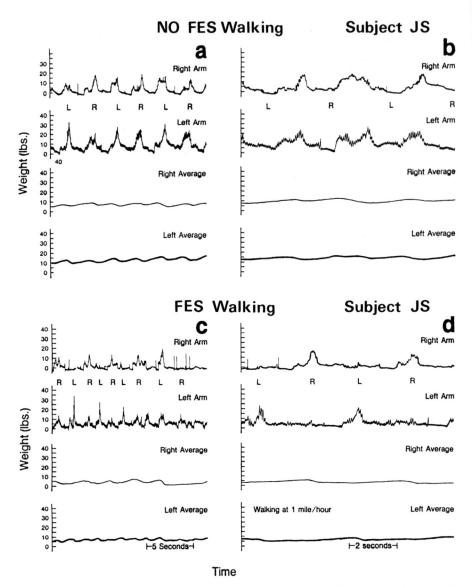

Figure 23. The weight borne on the hands during computer-controlled walking with and without electrical stimulation.

severe osteoporosis associated with spinal cord injury, this was a natural problem that had to be resolved so that product liability insurance could eventually be obtained on the products being developed. However, this particular system was too heavy, bulky, impossible to use successfully to go up and down steps, and impossible to use the toilet or get in and out of cars. Therefore, the system has been modified over the past few years from the standard 23 pound LSU reciprocating gait orthosis to a graphite system, as shown in *Figure 21*, with specific design changes on the hip and knee joints as shown in *Figure 22*. Using this combined system (total weight three and one half pounds) the

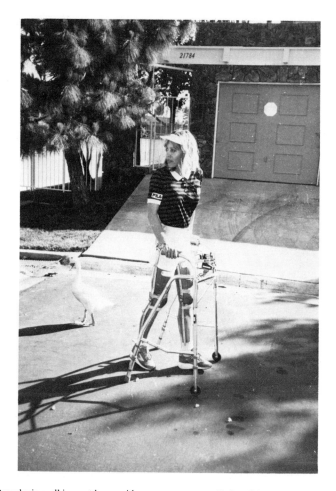

Figure 24. Paraplegic walking outdoors with a computer-controlled walking system.

oxygen consumption for walking has been reduced to that of voluntary exercise in non-paralysed individuals. Paralysed individuals have been able to walk distances well over five miles per day with no significant problems in terms of pressure points, muscle fatigue or cardiovascular stress. Although with conventional bracing the weight is borne through the arms, for computer-controlled movement via electrical stimulation of the legs, the weight on the arms during walking is almost insignificant. As shown in *Figure 23*, for example, in one paraplegic the weight shifted on the arms during walking is less than one percent of the body weight.

These systems have been tested extensively in the home and in places such as grocery stores as well as during walking outside in all types of weather (*Figure 24*). The future of these systems is quite promising, and certainly a great deal of research must be done before they can be brought to commercial use. The ability of such systems to be used outside the laboratory will depend on such a system being cosmetic, easy to use, reliable, providing good standing and sitting, and walking speeds in excess of one mile per hour.

The ideal system will be a fully implantable electrode and computer system but present technology has not, as yet, provided a material that can be implanted in the body for long periods of time without failure.

5. ACKNOWLEDGEMENTS

The author wishes to acknowledge the invaluable support of Janni Smith,R.N., who contributed her expertise,time, and encouragement to the ongoing research, and her photographic services and editing skills to this chapter. Warm thanks also to Karen Mueller for contributing her typing and editing skills.

6. REFERENCES

1. Goldsmith,H., Stewart,E., Chen,W.F., and Duckett,S. (1982) In *Application Intact Omentum to the Normal and Traumatized Spinal Cord.* C.Kao (ed), Raven Press, NY, USA.
2. Naftchi,N.F. (1982) *Science*, **217**, 1042.
3. Faden,A.I., Jacobs,T.P., Feuerstein,G. and Holaday,J.W. (1981) *Brain Res.*, **213**, 4515.
4. Perkins,C.S., Aguayo,A.J. and Bray,C.M. (1976) *Exp. Neurol.*, **71**, 515.
5. Kao,C., Bunge,R. and Reier,P. (1982) *Spinal Cord Reconstruction*, Raven Press, NY, USA.
6. Tenkoczy,A., Bajd,R. and Malez,C.M. (1976) *J. Biomech.*, **9**, 509.
7. Zealer,D.L. and Debo,H.H. (1977) *Trans. Am. Acad. Ophthalmol. Otolaryngol.*, **84**, 310.
8. Peckham,P.H., Mortimer,J.T. and Marsolais,D.B. (1976) *Clin. Orthop.*, **114**, 326.
9. Vodovnik,L., Crochetiere,W.J. and Reswick,J.B. (1967) *Med. Biol. Eng.*, **5**, 97.
10. Salomonov,M., Foster,J., Eldred,E. and Lyman,S. (1978) *Fed. Proc.*, **37**, 215.
11. Petrofsky,J.S. (1978) *Med. Biol. Eng. Comp.*, **16**, 302.
12. Petrofsky,J.S., Rinehart,J.S. and Lind,A.R. (1976) *Fed. Proc.*, **35**, 291.
13. Milner,M., Quanbury,A.O. and Basmajian,J. (1970) *Arch. Phys. Med. Rehabil.*, **51**, 540.
14. Scott,R.N. (1968) *Advances in Biomedical Engineering and Medical Physics.* Levine,S.N. (ed.) John Wiley and Sons, Inc., NY, USA, Vol. 2.
15. Reswick,J.B. and Vodovnik,L. (1967) *Artificial Limbs*, **11**, 5.
16. Crochetiere,W.J., Vodovnik,L. and Reswick,J.B. (1967) *Med. Biol. Eng.*,**5**, 11.
17. Lieberson,W.T., Homequest,H.J., Scott,D. and Dow,M. (1961) *Arch. Phys. Med. Rehabil.*, **42**, 101.
18. Petrofsky,J.S. and Phillips,C.A. (1983) *J. Neuro. Ortho. Surg.*, **4**, 153.
19. Petrofsky,J.S. and Phillips,C.A. (1979) *I.E.E.E. NAECON Record*, **79**, 198.
20. Petrofsky,J.S. (1979) *Med. Biol. Eng. Comp.*, **17**, 37.
21. Brown,W.P. (1973) *Neurol. Sci.*, **2**, 199.
22. Dubowitz,V. and Brooke,M. (1974) *Muscle Biopsy, A Modern Approach.* W.B. Saunders, Philadelphia, USA.
23. Van der Meulen,J.P., Peckham,P.H. and Mortimer,J.T. (1974) *Ann. NY Acad. Sci.*, **228**, 117.
24. Cherepakhin,M.D., Kakurin,L.I.,Ilina-Kakueva,E. *et al.* (1977) *Kosm. Biol. Aviakosm. Med.* **11**, 64.
25. Salmons,S. and Vrobova,C. (1969) *J. Physiol.*, **201**, 535.
26. Hudlicka,O., Brown,M., Cotter,M., Smith,M. and Vrobova,G. (1977) *Pfluegers Arch.,*, **37**, 141.
27. Brown,M.D., Cotter,M.A., Hudlicka,O. and Vrobova,G. (1976) *Pflueger's Arch.*,**36**, 241.
28. Cooper,E.B., Bumch,W.H. and Campa,J.F. (1973) *Surgical Forum*, **24**, 477.
29. Alam,M. and Smirk,F.H. (1983) *Clin.Sci.*,**3**, 247.
30. Lind,A.R., McNicol,G.W., Bruce,R.A., MacDonald,H.R., and Donald,K.W. (1968) *Clin. Sci.*, **35**, 45.
31. Petrofsky,J.S. and Lind,A.R. (1980) *Eur. J. Appl. Physiol.*, **44**, 223.
32. Krayenbuehl,H.P., Rutishauser,W., Schoenbeck,M. and Amende,I. (1973) *Am. J. Cardiol.*, **29**, 323.
33. Heltant,R.H., DeVilla,M.A. and Meister,S.G. (1971) *Circulation*, **44**, 982.
34. McCloskey,P.I., and Mitchell,J.H. (1972) *J. Physiol.*, **224**, 173.
35. Lind,A.R., Taylor,S.H., Humphreys,P.W., Kennelly,B.M. and Donald,K.W. (1964) *Clin. Sci.*, **27**, 229.
36. Petrofsky,J.S., Phillips,C.A. and Lind,A.R. (1981) *Circ. Res.*, **48**, 132.
37. Guttman,L. (1976) *Spinal Cord Injuries: Comprehensive Management and Research.* 2nd edition, Blackwell Scientific Publications, Oxford, UK.
38. Benassy,J. (1968) *Paraplegia*, **5**, 209.
39. Davis,G.M., Kofsky,P.R., Kelsey,J.C. and Shephard,R.C. (1981) *CMA Journal*, **125**, 1317.
40. Nilsson,S., Staff, P. and Pruett,E. (1975) *Scand. J. Rehabil. Med.*, **7**, 51.

41. Hjkeltnes,N. (1977) *Scand. J. Rehabil. Med.*, **9**, 107.
42. Clausen,J.P. (1977) *Physiol. Rev.*, **57**, 779.
43. Shephard,R.J. (1977) *Endurance Fitness.* 2nd edition, Univ. Toronto Press, Toronto, Canada.
44. Knutsen,E., Lewenhaupt-Olsson,E. and Torson,M. (1973) *Paraplegia*, **11**, 205.
45. Coutts,K.D., Rhodes,E.C. and McKenzie,D.C. (1983) *J. Appl. Physiol.: Respir. Environ. Exercise Physiol.*, **55**, 479.
46. Huang,C., McEachran,A., Kuhlmeier,K.V. and De Vivo,M.J. (1983) *Arch. Phys. Med. Rehabil.*, **64**, 518.
47. Bergofsky,E.H. (1964) *Ann. Internal Med.*, **61**, 435.
48. Axen, K. (1984) *J. Appl. Physiol. Respir. Environ. Exercise Physiol.*, **56**, 1099.
49. Claus-Walker,J. and Halstead,L. (1981) *Arch. Phys. Med. Rehabil.*, **61**, 595.
50. Downey,J.A., Chiodi,H.P. and Darling,R.C. (1967) *J. Appl. Physiol.*, **22**, 91.
51. Petrofsky,J.S. and Phillips,C.A. (1984) *CNS Trauma J.*, **1**, 57.
52. Pette,D., Ramirez,B. and Mueller,W. *et al.* (1975) *Pflueger's Arch.*, **361**, 2.
53. Astrand,P.O. and Rodahl,K. (1977) *Textbook of Work Physiology.* 2nd edition, McGraw Hill, NY, USA.
54. Simonsen,M., Foster,J., Eldred,E. and Lyman,J. (1978) *Med. Proc.*, **37**, 215.
55. Mueller,E.A. (1932) *Arbeits-physiologie*, **5**, 605.
56. Kralj,A. and Vodovnik,L. (1977) *J. Med. Eng. and Tech.*, **1**, 12.
57. Kralj,A., Bajd,T. and Turk,R. (1980) *Med. Prog. Technol.*, **7**, 3.
58. Kralj,A. and Jaeger,R.J. (1982) *Proceedings of the Fifth Annual Conference on Rehabilitation Engineering.* Houston, TX, USA.
59. Kralj,A. and Brobelnik,S. (1983) *Bull. Prosthet. Res.*, **10**, 75.
60. Holle,J., Gruber,H., Frey,M., Kern,H., Stohn,H. and Thoma,H. (1984) *Orthopedics*, **7**, 1145.
61. Brindley,G.,Polkey,C.E. and Rushton,D.N. (1978) *Paraplegia*, **16**, 428.
62. Marsolais,E.B. and Kobetic,R. (1982) *Proceedings of the Fifth Annual Conference on Rehabilitation Engineering.* Houston, TX, USA.
63. Petrofsky,J.S. (1981) *Eur. J. Appl. Physiol.*, **47**, 174.
64. Petrofsky,J.S. (1982) *Isometric Exercise and its Clinical Implications.* Charles C. Thomas, Springfield, USA.
65. Petrofsky,J.S., Danset,P. and Phillips,C.A. (1983) *Proceedings Third Int. Conference System Engrs.*, 213.
66. Petrofsky,J.S., Heaton,H.H. and Phillips,C.A. (1984) *Med. Biol. Eng. Comp.*, **22**, 221.
67. Petrofsky,J.S., Heaton,H.H. and Phillips,C.A. (1983) *J. Bioeng.*, **5**, 287.
68. Petrofsky,J.S., Heaton,H.H., Phillips,C.A. and Glaser,R.M (1983) *Collegiate Microcomputer*, **1**, 97.
69. Petrofsky,J.S. and Phillips,C.A. (1979) *Med. Biol. Eng. Comp.*, **17**, 583.
70. Petrofsky,J.S. and Phillips,C.A. (1982) *Fifth Annual Symposium on Neural Regeneration.* Proceedings National Spinal Cord Injury Foundation.
71. Petrofsky,J.S. and Phillips,C.A. (1983) *J. Neuro. Ortho. Surg.*, **4**, 165.
72. Petrofsky,J.S. and Phillips,C.A. (1985) *CRC Crit. Rev. in Biomed. Eng.*.
73. Petrofsky,J.S., Phillips,C.A., Douglas,R. and Larson,P. (1985) *J. Neuro. Ortho. Surg.*, **6**, 230.
74. Petrofsky,J.S., Phillips,C.A. and Heaton,H.H. (1984) *Comp. Biol. Med.*, **14**, 135.
75. Petrofsky,J.S., Phillips,C.A., Heaton,H.H. and Glaser,R.M. (1984) *J. Clin. Eng.*, **9**, 13.
76. Petrofsky,J.S., Phillips,C.A. and Hendershot,D. (1985) *J. Neurol. Ortho. Med. Surg.*, **6**, 190.
77. Petrofsky,J.S., Phillips,C.A., Larson,P. and Douglas,R. (1985) *J. Neurol. Ortho. Med. Surg.*, **6**, 219.
78. Phillips,C.A. and Petrofsky,J.S. (1986) *J. Neurol. Ortho. Med. Surg.*, **7**, 225.
79. Petrofsky,J.S. (1987) *Memorial Lectures for the Late Dr Yutaka Nakamura.* Japan: Tamikazu Amako, Japan.

CHAPTER 8

Simulation techniques for teaching

JAMES E. RANDALL

1. INTRODUCTION

During the first decade of microcomputer history the bulk of the physiological teaching software has been simulations of specific phenomena, such as respiratory mechanics, acid-base balance, axon action potential and cardiovascular mechanics. These programs are generally written in interpreted BASIC and are of limited size (1). Meanwhile the memory capacity and other hardware embellishments have outgrown the capabilities of the original BASIC interpreters supplied with virtually all microcomputers.

This chapter focuses upon the technical details of realizing the full hardware capacity for physiological simulations using modern microcomputers. Recent BASIC compilers that are compatible with the familiar interpreters allow better utilization of the hardware. No longer limited to 64 K of memory (RAM) the programs can be both larger and easier to write. The particular emphasis here is upon the structured nature and modularity of the source code. This approach has long-range accumulative benefits with the convenience of user libraries containing standardized utility routines and banks of physiological functions.

Supporting material includes necessarily subjective comments on the hardware and software being used for teaching in university physiology departments. The programming examples are given for the popular MS-DOS (IBM-PC) machines and the QuickBASIC compiler, but the ideas have broader application. The single most important ingredient for successful use of microcomputer simulation still is the participation of teachers who are enthusiastic about the approach.

2. PHYSIOLOGICAL SIMULATION

Personal microcomputers are having considerable impact upon education. The low cost of the hardware and the convenience of programming and system operation provide a creative outlet which stimulates innovative approaches in teaching. Secondary schools are meeting the challenge of providing universal computer literacy, but more specialized upper level courses have been slower to take advantage of this opportunity. Word processors are replacing typewriters for preparing manuscripts and class materials but there are few individuals motivated to spend time with the technical details of producing teaching software.

Of all the biomedical sciences physiology is the discipline in the best position to take advantage of the computational power of a microcomputer (2). The conversion from vitalistic to mechanistic explanations of body function drew heavily from mechanical models and analogies with mathematical models as a natural extension. Insight into

Simulation techniques for teaching

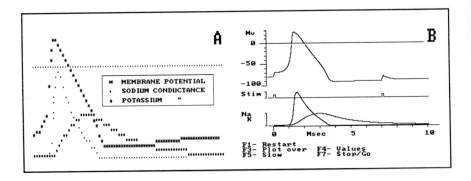

Figure 1. The Hodgkin and Huxley simulation of axon excitation as plotted in two simulation eras. **(A)** Mainframe line printers were often used to produce graphics having limited resolution. **(B)** Today's microcomputers typically control 640 × 200 individual pixels on the screen. The new generation of video monitors modulate red, green and blue beam intensities to provide a wide range of colours that add pictorial realism.

complex systems depends upon formal models which extract the essential characteristics in a digestible form.

Mathematical simulations, now practical with microcomputers, permit students to explore the overall consequences of changing a single parameter in a model. Graphic displays provide a unique method of appreciating the dynamic nature of homeostatic compensations with a range of time constants of seconds for neural responses to days for fluid distributions. *Figure 1* contrasts earlier line printer graphs with those for current medium-resolution graphics for a popular teaching simulation. Here a student picks the amplitude and duration of an electrical stimulus and sees the membrane potential and conductances as calculated by the Hodgkin and Huxley equations.

The development of simulations sharpens quantitative skills, providing experimental approaches to non-linear formulations not possible by traditional analytical methods. The explicitness required clarifies the thinking, advancing physiology as a scientific discipline. Furthermore, simulation may play a vital role in furthering the integrative nature of physiology in an era marked by advances in molecular biology, neural science, and biophysics.

Interpreted BASIC has been the common programming language for nonprofessionals, referred to here as casual programmers. Since this form of BASIC had its origins in an era when personal computer memories were measured in increments of 4 Kbytes it is very limited in scope. Though serious modellers use other languages, as mentioned in Section 4, these each have their own characteristics which require repeated use to maintain any level of programming skill.

It is the thesis of this chapter that newer BASIC compilers, of which there are several, can now provide most physiologists with a tool for realizing the power of their microcomputers for teaching applications while working with a familiar programming language. Besides allowing larger programs this form of BASIC now allows structure and library modules in a friendly environment with advanced debugging, all features that have increasing importance as the applications grow in size.

2.1 Why acquire programming skills?

The modest cost of today's microcomputers conceals their tremendous computational power and versatility. For the person working with word processors the system is an expansion of the familiar typewriter. For the laboratory investigator the computer replaces banks of relays and/or strip chart recorders used to control experiments and acquire data. But for teachers the small computer opens new vistas which transcend the familiar modes of instruction. For this application we still are searching for prototypes, such as interactive tutorials and testing, control of video images and image analysis including anatomical serial sections and molecular structures.

For mathematical simulation teaching applications we are now in an era analogous to the development of word processors when they were merely text editors. During these exploratory attempts interaction with a programming language is necessary. Three different levels of involvement are possible.

At the most sophisticated level someone has to translate a physiological process or phenomenon from a quantitative model into a computer program or module. Iterative methods, practical with tireless computers, provide methods for keeping track of numerous non-linear rate equations describing body function. A classic example of this is illustrated in *Figure 1*. The larger models are often produced by specialists using languages which discourage widespread application by others. On the other hand, effective teaching requires knowledge of just what is and what is not in a given model. A more powerful BASIC may encourage casual programmers to tackle models for their speciality areas and for such models to be more widely distributed and evaluated.

If various organ and body functions can be compartmentalized into modules developed by research specialists then teachers might be able to collate these into a variety of instructional formats, for example tutorials, simulated clinical cases, or experiments. Here the programming involved is one of calling upon various subroutines; for example pulling out specific variables to be plotted according to a generalized graphics subroutine. Section 9.5.4 presents a flexible scheme for doing this by preparing text files which are run on previously compiled simulation software.

The continuous proliferation of textbooks in stable disciplines, such as statistics, gives testimony to the fact that there are numerous approaches and styles to teaching a subject. Thus there will be a dissatisfaction with the emphasis or organization of a particular simulation package as provided by the mechanism of the previous paragraph. The knowledgeable user may be able to understand the limitations of a particular model and to make minor changes in the manner in which one interacts with it for teaching purposes.

Thus the pattern could involve the research specialist who develops the modules, the organizer who combines these into a teaching program, and the final user who makes minor adjustments to meet local objectives and personal taste. These levels are generally applicable to other teaching applications, not just to simulations. However, this scheme presupposes that all materials are non-proprietary and thus in the public domain, in the spirit of research publications. Simulation software is now generally freely exchanged but this may not be the case if commercial interests begin to prevail as is the case with spreadsheets and word processors.

3. HARDWARE

In contrast to the dedicated hardware configurations of laboratory microcomputers the needs for teaching simulations must be able to accommodate software available from a variety of sources. The rapid advances in memory, graphic displays, manufacturing and marketing make it impossible to make a thorough, objective evaluation of all architectures. Usually selection is based upon local economic factors and past personal experiences. Fortunately the leaders, the classic Apple II, the Macintosh and IBM-compatibles, have retained some degree of upward compatibility so that new designs will run old software. But it also means that numerous earlier installations may not be able to take advantage of evolutionary changes.

Hardware configurations generally emphasize one of two applications: student use or program development. Because of their number, student work stations are generally minimum configurations chosen to accommodate currently existing software. These tend to be conservative designs without the extras costs of numerical processors, 20 megabyte (Mbyte) data storage or laser printers. On the other hand, installations used for writing software do require aids which allow the time of professionals to be used productively. In particular, an inexpensive hard disc is essential for compiling the kinds of programs being discussed here. Thus a person could use a new Mac SE model to support numerous older Mac's.

3.1 Architectures

It is my experience that the classic Apple II series and the BBC microcomputers are still very popular in the secondary schools but that the IBM PC (and compatibles) prevail in the professional schools where physiology is frequently based. Now IBM is phasing out the PC line, replacing it with their IBM Personal System/2 (PS/2) line. The major orientation for the top of the line models is for the corporation environment rather than for the individual. The Model 25, promoted for educational use, is actually an upgrading of the original PC.

With a user-base of some ten million MS/PC-DOS computers from a variety of manufacturers the original PC architecture seems likely to continue in popularity for teaching and other academic applications.

There are currently approximately one million Apple Macintosh microcomputers in use. This architecture has growing popularity in teaching applications because of the ease of its use, its outstanding graphics and favourable academic pricing. Its future seems assured by the fact that the new models, the SE and the II, reverse the original closed-architecture concept while building upon existing models. Widespread use of this kind of microcomputer has been hampered by Apple's tight control of the market, preventing inexpensive clones.

3.1.1 *New developments*

The new microcomputer designs may not have the impact on the typical teaching application that they will in the corporate world. The present 640 K memory limit of the IBM PC and Model 25 provides plenty of room for contemporary teaching simulations. There are more economical ways to increase computation speed for existing software than by replacing the whole computer system.

Many of the graphics improvements, as for the Atari, Amiga, Apple II GS and IBM PS/2 models involve a marked departure from traditional CRT circuitry, and thus are more expensive and diversified. Specifically these increase the number of colours and levels of white by analogue modulation of the beam intensity whereas older methods were either off or on at one or two levels of intensity for the red, blue and green beams. These changes, along with increases in resolution, can be put to effective use, but they are not compatible with the readily available monitors and video projectors which are the common standards in teaching environments.

The thrust of the larger IBM and Macintosh machines will be appreciated where several users need to share common information sources or a given machine might be able to perform a number of independent tasks simultaneously. The operating systems are comprehensive, expensive and take an enormous amount of memory. While useful for program development they are not apt to become the norm for individual students. My personal prejudice is against having several students share a single microcomputer using separate terminals. To me, this is a regression back to the old mainframe era which gave users a sense of being driven remotely rather than being in control of a small box.

Thus at the time of writing my perception of what is practical is to show how to realize the potential of available popular microcomputers.

3.2 The IBM-PC

The modular software theme of this chapter is discussed in terms of the original IBM Personal Computer and its many near compatibles. My recent experience has been with this configuration and the vast majority of worldwide software inquiries convince me that this is currently a standard apt to be available for the majority of the readers of this volume. In particular, the newer structured BASIC compilers required for modularity are widely available. As indicated above, I am very supportive of the Apple Macintosh computers for teaching simulations, but sense that the developmental software has emphasized their graphics capacity rather than numerical computations.

A major contribution from IBM was the establishment of a *de facto* standard for hardware, a disc operating system, and interpreter BASIC. It is estimated that there are at least ten million of these configurations in existence selling in the US for as little as $1000, including monitor and printer. Typically the clones come with 640 K of memory (RAM), one 5.25 in disc with 360 Kbytes of capacity, and monitor/printer interfacing. These usually include several slots to accommodate a wide variety of hardware embellishments, including hard discs, modems, and lab data acquisition.

The IBM AT, noted for its upgrading of the original 8088 to the more powerful 80286 microprocessor, runs original PC software at a faster speed. New serious individual users are apt to move to the PS/2 models while economical student work stations are likely to use clones, or the Model 25 if suitably priced. A number of so-called accelerator boards exist which can enhance existing PC's if speed is the objective. For the main purposes of this chapter the PC, XT, AT, Model 25 and their compatible equivalents are considered together.

3.2.1 *Graphics interfaces*

The original IBM PC had two different graphics interface boards. The Monochrome, meant for business applications, was limited to text with some graphic characters suited

Simulation techniques for teaching

Table 1. IBM personal computer display adapters.

		Monitors
MDA	Monochrome Display Adapter Text and ECS symbols	Digital, fast scan
Hercules	Added 720 × 348 pixel graphics to MDA Not supported by all BASICS	Digital, fast scan
CGA	Colour Graphics Adapter 16-colour text and two graphics modes 320 × 200 pixels with 4 colours 640 × 200 pixels with 2 colours	Digital, standard video scan
EGA	Extended Graphics Adapter Supports MDA and CGA, adds 640 × 200 pixels with 16 colours 640 × 350 pixels with 16 colours	Digital, multi-scan
MCGA	Multi Colour Graphics Array Subset of VGA for Models 25 and 30 CGA 4-colours increased to 256 Improved text resolution with shading Requires adapter for EGA graphics	Analogue, fast scan
VGA	Video Graphics Adapter Standard for Models 50, 60, 80 640 × 480 graphics with 256 colours High-resolution text	Analogue, fast scan

for rectangular boxes. The Colour Graphics Adapter (CGA), of more interest in teaching simulations, supplies four colours at 320 × 200 pixel (display element) resolution or two colours at 640 × 200 resolution. Note that the monitor electrical characteristics for the Monochrome and CGA cards are different. The CGA is currently the most widely used display in simulation courseware, but there are alternatives to complicate the picture (see *Table 1*).

The most serious source of PC incompatibility among products from different manufacturers is in the support for graphics. Because the predominant business application is for text and not coloured graphics many of the inexpensive models do not support the CGA required by most simulation software. The monitor specifications may be like those for the IBM Monochrome rather than those of the more standard scan rates found in classroom display units. Thus there may be an extra expense in acquiring a suitable interface card and/or monitor. The Hercules Graphics Card established a secondary graphics standard for the IBM Monochrome but it cannot be addressed by the standard BASIC interpreters or compilers.

A further complication is that IBM followed with an improved graphics standard known as Enhanced Graphics Adapter (EGA). This supports older CGA software and popular monitors, but when used with newer, more expensive monitors can display 16 colours at the 640 × 200 resolution or even at an improved 640 horizontal × 350 vertical pixels. Newer versions of the BASIC interpreters and compilers provide software access to these graphics by adding commands SCREEN 7–10.

Extensive advertising for both EGA cards and EGA monitors has suggested that this format was becoming a standard, replacing the modest capabilities of the CGA. Now the displays with the new IBM PS/2 system (MultiColor Graphics Array for Models 25 and 30, Video Graphics Array for Models 50–80) introduce a new variable, modulation of beam intensity. The attribute for each bit is stored in 8-bits in RAM. This provides 256 colours out of a total possible 256 000 different colour combinations provided a proper monitor is available. The new design is an improvement over EGA but current models do not support existing EGA software without special hardware.

There are numerous alternative sources for both display adapters and video monitors which accommodate these *de facto* standards from IBM.

My own conservative approach at this time is to stick with the original CGA standard available at most installations using the PC computers. But, since the purpose of this discussion is to encourage modularity it should be practical to adapt graphics modules to changing hardware.

3.2.2 Numerical coprocessors

The PC class of microcomputers generally come with a socket in parallel with the main microprocessor, an 8087 for the 8088/8088 or an 80287 for the 80286. With matched software the 8087 provides improved numerical computation speed and precision because multiple software operations are replaced with a single command recognized by the coprocessor.

The original BASIC interpreters do not accommodate the 8087 numerical coprocessor but most compilers do. The general idea is that numerical computations are handled internally at 80-bit precision and that the evaluation of transcendental functions do not require time-consuming series expansions by software. The disadvantage is that numerical data are stored in memory in groups of 8-bit bytes so that time is spent in getting stored data in and out of the numerical processor. The computer software must be optimized in order to realize maximum benefits.

In my own software experience the improvements in computational precision have not been important. The major speed improvement occurs with programs requiring the evaluation of a large number of exponentials, as for the Hodgkin and Huxley model for axon excitation. Otherwise, for most numerical computations the overall speed improvement is usually 2.5–3 times.

3.3 Use in lecture halls

There is a tendency to think of the application of a microcomputer as an aid for teaching as one-on-one with a single student. While this is the most common situation, two new developments will contribute to greater use of this technology in the lecture hall and for demonstrations to large groups. Laptop microcomputers now have the full capabilities of desk top models, but can be easily transported, weighing no more than a large textbook. These are supplied with interfacing for the large display video monitors found in some classrooms. Text legibility for large groups requires that programs use 40 characters/line displays.

Simulation techniques for teaching

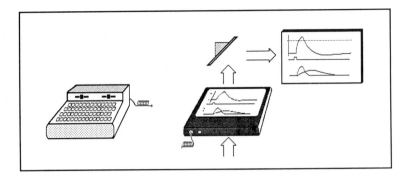

Figure 2. LCD plate which displays microcomputer output on conventional overhead projectors. Similar to the displays on laptop computers, the device plugs into the video output and is used as though a transparency.

Another useful device displays computer output onto a large wall screen via a standard overhead projector. Instead of using static transparency films a liquid crystal display (LCD) is connected to the video output of the microcomputer. Computer text and graphics is diverted from the computer screen, an internal LCD for laptops, to an external LCD placed on the projector. This external LCD appears similar to a transparent picture frame connected by cable to the computer and to a power source (see *Figure 2*). Both LCD and laptop computer easily fit into a briefcase and can be carried across campus.

Attractive as portability is, LCD technology is still in its infancy. What appear as assorted colours on video monitors show as various intensities of opaqueness for the LCD. The contrast between displayed material and the background is poor, though a newer 'super-twist' LCD is an improvement. For high-wattage projectors the heating of the crystals reduces the contrast. Some devices cool the LCD with a fan.

Interfacing varies with the manufacturers. Typical input is required to be from a conventional red-green-blue (RGB) connector as contained on IBM CGA or equivalent. A variation is the ability to accept composite outputs, as from the classic Apple II series. Another requires a special internal interface board and is impractical for laptop computers.

One disadvantage for using the Macintosh in teaching situations is that its monitor's design is matched very specifically to the computer so that the Mac cannot utilize conventional large video monitors typically used in classrooms. Projection monitors and LCD devices are available for Mac computers but these can not be used with conventional video cassette recorders used in some courses.

I find the convenience of the laptop with permanent classroom video monitors conducive to making use of dynamic simulations during lectures. The action potential demonstrations hold student attention and provide an introduction to the operations they will perform on their own. Other effective examples include the convergence of summed statistical events, for example the Gaussian distribution of sample means and the summation of axon single-channel conductances. Because of the poor contrast of the original LCD overhead projector devices, those displays are received less enthusiastically.

4. SOFTWARE

Since the production and testing of teaching courseware is so time-consuming there is always interest in finding efficient programming languages to speed the task. Two viewpoints have prevailed, each with their proponents and detractors. The general-purpose languages, FORTRAN, BASIC, Pascal and C, give maximum flexibility and can tap the full computational power of the microcomputer. Special-purpose languages, such as Pilot and Logo, involve much simpler programming steps which limit the range of possible operations. Authoring systems place a shell around such an authoring language to assist instructors create the steps in a lesson interactively.

Approaches typified by Pilot may be useful for some topics, such as language drills, but they are too slow and too limited to handle the complexities of multi-variable mathematical simulations of the type under discussion here. Several arguments can be made that for physiological teaching simulations some variation of a general-purpose language is proper.

A given language may be more proficient in the execution either of simulations or of tutorials. In very general terms, for simulation the computer asks the operator for conditions whose consequences are then shown in either numerical or graphic form. In tutorials the computer asks the questions and the student supplies the answers. The former may require computational speed while the latter requires efficient handling of text.

4.1 Simulation languages

Simulation languages interject a layer of simplified operational commands around a general-purpose language so that programmers need never see the technical details. In fact the same simulation language might be written in FORTRAN, C or Pascal so that improvements in performance at that level are transparent to the persons developing the simulations.

Large-scale simulations used in the design of aircraft, rockets and satellites involve dozens of programmers working on specialized aspects more or less independently. The languages developed for these kinds of applications have numerous versions adapted for microcomputers. These are useful for graphic displays of dynamic solutions but lack the extensive text capabilities required for teaching use.

Although not a language SCoP is another example of a way to observe dynamic solutions without the burden of excessive programming details. SCoP (Simulation Control Program) is a general-purpose interactive simulation program developed at the National Biomedical Simulation Resource located at Duke University in Durham, North Carolina. It is written in C to facilitate interactive study of biological systems which can be represented by algebraic and differential equations. Modellers set up their equations including calls to plotting routines according to a standard format. This source code is then compiled and linked to the SCoP library. The program, with several physiological examples, is available for IBM PC computers at a modest handling charge.

4.2 Drivers for text files

Another technique for programming efficiency is to have the instructor build ASCII text files using familiar word processors or editors. These files, written for a specific lesson, are then used with previously compiled programs, here called drivers, to perform a simulation or present a tutorial. This approach taps the power and speed of the computer while letting the instructor focus on the lesson.

MacAid is such a driver program. It acts as an interpreter for text files and provides capabilities to integrate instruction, multiple-choice questions, assessment scoring, and instructor-defined branching logic. It was designed for use with the physiological models MacMan, MacPuf (3), MacPee and MacDope, obtainable from IRL Press Ltd, Eynsham, Oxford, UK.

A modest example is given in Section 9.5.4 of this chapter as a convenient way to prepare dynamic simulations of a large comprehensive model but in which the attention is being focused upon only a few relevant variables. A text editor is used to prepare a file containing the names of the desired variables and their scales. At class time the driver reads that file and runs the model. This combines flexibility with maximum computational power without having to be bothered with the time-consuming programming.

4.3 General-purpose languages

Since physiology does have its quantitative aspects, any familiarity with a general programming language will have broader application than just for teaching. It is quite likely that the teacher also may need to know how to use the microcomputer to control a laboratory instrument or perform a unique kind of data analysis. The modular structure of Pascal, C and BASIC compilers make programs written in these languages much easier to read and to maintain as compared to the original FORTRAN compilers and BASIC interpreters.

4.3.1 *Alternatives to BASIC*

When FORTRAN is used for microcomputer applications it is generally in order to take advantage of existing mainframe software or because the applications arise in an environment that has been using computers for quite some time. Student difficulties with mainframe FORTRAN in the 1960's prompted Kurtz and Kemeny to design BASIC as a simpler beginner's language. It is particulary convenient for text strings. When Allen and Gates wrote a BASIC interpreter for the 4K Altair microcomputer it was very limited out of necessity, but it set a pattern which is still being followed for interpreters, as described below.

As application programs grow in size and complexity they involve the cooperative efforts of many individuals. Readability and maintenance of programs then becomes a significant factor in the overall productivity of a project. FORTRAN and BASIC code can easily combine numerous GOTO statements to produce convoluted code not recognizable even to the original author. In response, Wirth developed Pascal as a highly structured language which placed stringent demands upon the programmer to work with procedures and to define data types before a given variable was used. Such discipline is good training for programming in any language so it is used widely in university courses.

Borland's Turbo Pascal brought that language to microcomputers, with over half a million copies being sold for use on the IBM PC, Macintosh, and older 8-bit machines.

Besides its structure and data types its popularity results from the combined editor and compiler as a single integrated unit which becomes interactive.

The language C also is structured and requires data typing. It most often is used by professional programmers because it works closely with the hardware, making it possible to get maximum performance at nearly machine code level, but, it places considerable responsibility upon the programmer to avoid errors. Recent developments include Microsoft's QuickC and Borland's Turbo C which follow the tradition of Turbo Pascal regarding interactive programming.

For the past 30 years I have watched this evolution from FORTRAN to Pascal to C from the vantage point of the physiologist. Productive use of these languages requires enough use to keep refreshing their various nuances and syntax requirements. Many physiologists use one of these as a primary computer language. I am convinced that there is a much larger number of biomedical professionals who grasp the simplicity of BASIC but who have insufficient contact with the other languages to develop any familiarity. For this reason I am enthusiastic about the potential of the new BASIC compilers which overcome many of the shortcomings of the popular interpreters.

4.3.2 BASIC interpreters

Most microcomputers are supplied with some variation of the BASIC interpreter originally written for the Altair by Paul Allen and William Gates, founders of Microsoft. No doubt the convenience of that version of the language, limited as it was, had considerable bearing on the rapid incorporation of small computers into all kinds of applications. Different versions of interpreters have been supplied with specific microcomputer models from a variety of manufacturers including Apple, Commodore, Radio Shack, Zenith and IBM. Since a given interpreter was closely matched to specific hardware the language took full advantage of the features of that machine, including graphics and sound. This was not the case for languages which had to fit a wide variety of computer configurations.

This section gives an overview of the interpreters which come with the IBM PC and for those supplied by Microsoft for compatible machines. The latter are often called GWBASIC. These interpreters take up 64 K of RAM and allow another 64 K for user programs and data. This limits both the features of the interpreter and the size of programs. The utility of the design arises from the fact that syntax errors halt the program execution giving debugging help. However, this is done at the expense of execution speed because the interpreter changes each line of source text code into machine code no matter how many times it reads that line, as during a repeating loop.

Interpreted BASIC source code must identify each line with a number. The general structure is to follow these numbers in sequence unless there is a branching. Thus a program might branch to either of two line numbers depending upon the answer to a question. Operations which are used repeatedly may be placed in modular subroutines and be called as needed. As programs grow in size and complexity the use of numbers as references becomes cryptic, whereas appropriate names make the operations obvious.

Another disadvantage of the interpreter is that variables are global throughout the program. Changing the value of a variable at any one point in the program changes it everywhere else, including all subroutines. This discourages the use of subroutine libraries in which the action of one routine might inadvertently influence another that happened to have variables of the same name. Large programs need modularity and the modules must be isolated from one another except for well-defined methods of passing specific variables.

There are several other restrictions of the BASIC interpreter which are overcome by the newer compilers mentioned below.

4.3.3 *BASIC compilers*

The original Apple II had a 10 K interpreter permanently located in memory (ROM) and 16 K of RAM available for programs and storage of graphics. Once a program was debugged using the interpreter its source code could be passed through a compiler which changed *all* statements to machine code in one pass. The resulting compiled code was recognized by the microprocessor and could run much faster. Thus one of the objectives of compilers always has been to increase execution speed. Modern compilers do much more than that.

The first PC BASIC compilers, obtained from IBM and from Microsoft, were used primarily for increasing execution speed. However, the code produced by the compiler, known as a relocatable object code, is in a form that can adapt to any location within memory. The Microsoft LINK program is then used to combine the user's object code with specific addresses from a library of subroutines to produce the final executable program.

As the popularity of the PC computers grew BASIC compilers became available from a variety of sources (see *Table 2*). Besides increasing speed, these corrected many of the deficiencies of the *de facto* standard established by the widespread use of the microcomputer interpreter. True BASIC saw the return of the originators of mainframe BASIC and their product incorporates many of the mainframe structural features bypassed in the first 4 K interpreter. It is particularly noteworthy because of its ability

Table 2. BASIC compilers.

Better BASIC	Summit Software Technology PO Box 99, Babson Park Wellesley, MA 02157
QuickBASIC	Microsoft Corporation PO Box 97017 Redmond, WA 98073-9717
True BASIC	Addison-Wesley Publishing Co. Reading, MA 01876
Turbo BASIC	Borland International, Inc. 4585 Scotts Valley Drive Scotts Valley, CA 95066
ZBASIC	Zedcor, Inc. 4500 E. Speedway Blvd. Tuscon, AZ 85712-5305

to handle a wide variety of different graphic displays with the same software.

The source code for ZBASIC is easily transportable between machines having different architectures. Better BASIC is another version which has its own attributes and there are others. Their commands depart from the established Microsoft interpreter to varying degrees and the total cost of a fully implemented package which can be distributed to several users may be relatively expensive.

4.3.4 QuickBASIC versus Turbo BASIC

The programming convenience of Turbo Pascal stimulated the more recent compilers to become interactive. Two are particularly popular. Microsoft's QuickBASIC (QB), version 3.0, and Borland's Turbo BASIC (TB) are being heavily promoted in the hope that one will become the standard which replaces the Microsoft interpreter.

Both allow large programs, support the 8087 numerical processor and add new structural features which enhance the modularity required for large programs. By default variables are confined locally to subprograms with several options for sharing them. Equally important, they allow compiling in an interactive mode which closely couples the editor with the compiler, almost the same as an interpreter. Also, error messages now say 'Missing Right Parenthesis' rather than simply 'Syntax Error'.

The differences between the two versions may not compromise the final programs produced, but do influence their respective conveniences for programmers. Program execution speeds for machines containing the 8087 are comparable for the maximum numerical precision. However, in machines without this numerical chip there is a marked difference in computation speed, at least for current compilers. When QB and TB emulate the 80-bit precision of the 8087 both compilers produce programs which are excessively slow for teaching applications which involve computations. QuickBASIC 3.0 comes with an alternative compiler which produces faster programs with only 32-bit precision, the same as that of the interpreter. Thus there is a significant speed advantage for QB over TB version 1 for machines without the 8087.

There are several advantages of QuickBASIC for program development. It allows the user to prepare precompiled subprograms, placing them in a memory-resident library while editing and compiling related modules which use those subprograms. This provides considerable saving in time to correct errors during compiling. On the other hand, TB requires that the compiler search through the source code for both the portion under development plus all subprograms. Though the subprograms may be included from disc files, the fact that these files can not be nested diminishes the concept of modularity. The TB is compromised further by the fact that it aborts compiling upon finding the first error rather than keeping track of as many as 26 errors as does QB 3.0.

QuickBASIC contains an unusually effective debugger accessible for the code currently being compiled, but not that in the user's library. Several breakpoints may be inserted to pause the program at specific steps. At that time any computed variable may be examined while any one may be followed as the program executes single steps. This is particularly helpful for following branching within the program and for discovering what some subroutine has done to a variable.

There are slight differences in BASIC command extensions which the two compilers do not share. Since QB modules may be separately compiled while TBs can not, one source code may not necessarily serve the two compilers without slight changes. Turbo

BASIC allows recursive calls to a given subroutine, supports the newer IBM PS/2 graphics and has the better editor.

From my own experience QuickBASIC's programming conveniences outweigh those of Turbo BASIC. Therefore, QB is the version used in the following examples.

5. USING THE COMPILER

One of the benefits of using the QuickBASIC compiler for program development is that compiled modules can be combined into a memory resident user's library. Thus during the debugging of one program segment only that portion's source code needs to be scanned. This is much faster than the two alternatives of (i) compiling the entire program including user subroutines which are in source code form or (ii) the traditional formation of compiled object code files which are then linked in a separate stage without interaction with the editor.

I have found that the convenience of the QB approach encourages me to make minor cosmetic changes in very large programs which otherwise would be too onerous. Though not as fast as interpreter debugging the interactive environment is a marked contrast to the batch process methods which took as much as 15 min for some large FORTRAN programs.

Since the purpose of this chapter is to stress a modular design for production of teaching simulations this section contrasts three possible methods of compiling. Note that effective use of QuickBASIC for large programs requires at least 512 K of memory and a mass storage device, both of which are now reasonably priced.

The QB compiler can be implemented in two different ways. One is as a DOS command line which names the source code file which is to be compiled, along with one or more of several optional switch settings. This bypasses the editor and the attractive interactive debugging features. The other more useful way to start either the QB or TB compilers is to enter these program names at the DOS command. This places you into the editor where various options can be called interactively.

5.1 Linking compiled program modules

The traditional method of combining several subprogram modules into a single executable program is to compile each module separately and then to use the Microsoft linking program (LINK.EXE) to combine these into a single unit. As shown in *Figure 3B* the linker combines both user-compiled object code files and a BASIC library. Although this is a required final step for the approach of Section 5.3 it is a time-consuming procedure during the preliminary debugging stages of program development.

Consider an example in which one main object file and three separate subroutine files have been compiled, either using the interactive or the batch approach. Their respective object code files then can be linked by the single DOS command

C> LINK MAIN.OBJ + SUB1.OBJ + SUB2.OBJ + SUB3.OBJ;

The resulting output file would be called MAIN.EXE and would either incorporate a BASIC library into the program or use a separate library loaded at run time.

5.2 Compiling to executable code

Both Quick- and Turbo BASIC compilers have an option, set from a menu while editing the text source, which produces an executable program in a disc file that can be run

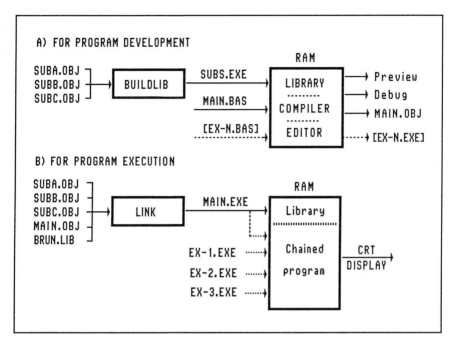

Figure 3. Compiling (**A**) and using (**B**) programs which share a common library of subroutines. Interaction during compiling and debugging is improved by having the library memory resident. Executed programs can CHAIN between different routines, leaving the library in memory. This conserves disc space and speeds loading time. Here EX-1, EX-2 and EX-3 are different exercises which use the subroutines loaded into memory by MAIN which could be no more than a menu showing available programs.

from DOS. This has the advantage of interactive debugging and produces files with an extension of .EXE which do not have to be linked. For TB the output is a complete self-contained unit which requires no library support. The QB output may depend upon a runtime library or it may be chained from a main program which contains the required library and subroutine support.

The ability to chain to small program segments while retaining large underlying support resident in memory has advantages in conserving disc space and loading time. As an example, a large comprehensive physiological model, to be described later, along with numerous utility subroutines are compiled as a program called MENU. This program is over 100 Kbytes in size. Selections from the menu then chain to one of several tailored lessons compiled to .EXE files taking about 10 Kbytes each. These all have access to the various subroutines without the need to have several 100 K files with duplicated material (See *Figure 3B*). Turbo BASIC does not support chaining and requires the self-contained program files.

5.3 Compiling to memory

Modern BASIC compilers are designed to take advantage of the large amounts of memory now present in microcomputers. A typical sequence starts the compiler, uses its editor to enter the source code, then compiles directly to memory. QuickBASIC accumulates any syntax errors and then returns to the editor for their sequential correction. Typical compiling time for a 10 K program is about 10 sec.

At the end of compilation program execution starts immediately while control remains within the editor. Options include the ability to single step through the program and also set break points where the program will pause during execution. At these pauses the value of any variable can be obtained and/or one value may be watched during program execution. Once the program is operating satisfactorily it can be compiled to a disc file either as a self-contained program or as an object code to be linked with other subprograms.

5.3.1 *User's library*

Programming efficiency is achieved if the program can be broken into subprograms called from a main control module. The individual components can be written, debugged and saved as object files one at a time. A special feature of QB is that a number of these previously compiled subprograms can be combined into a single user's library which then is available for inclusion as utilities in much larger programs. Following sections give examples of such subprograms.

Suppose that three individual subprogram modules had been designed and tested as separate entities and that these were compiled and saved as relocatable object files (see *Figure 3A*). These can be combined with the supplied BUILDLIB.EXE program as a DOS command

C: > BUILDLIB SUB1.OBJ SUB2.OBJ SUB3.OBJ, SUBS.EXE

The user's library output file would then be named SUBS.EXE. While writing a MAIN program which calls subroutines from this library the latter may be placed in memory where it will be instantly available. This is done by starting the QB program with a special flag followed by the full name of the library, including the .EXE extension.

C: > QB/L SUBS.EXE

The editor loads as before but memory contains the various subroutines rapidly accessible without further compiling or linking. One then can proceed to compose and interactively compile and correct the MAIN program. Calls to the library subroutines are handled automatically, but these sections of code can not be interactively debugged as can the MAIN program. Once the current program is functioning it can be saved as an object code file on disc. Then, as described in Section 5.1, the debugged object files are linked into an executable program.

Remaining sections give illustrations of functional subprograms and how these can be combined for teaching simulations. Once the subprograms are written and tested they may be combined into specific libraries for further use with a minimum of programming effort.

6. INTERACTIVE TECHNIQUES

Regardless of how valid the physiological model, or how clever its software coding, or how colourful the display, a teaching simulation which leaves the student confused about what to do next will be a failure. The subtleties of effective teaching, still being learned for computer instruction, require skills that differ considerably from those required for mathematical modeling and computer programming.

Nevertheless, programming style can have significant influence upon the final product. Applicable principles involve simplicity, clarity, consistency and responsiveness. This section outlines specific examples of these which can be incorporated as the programs are being designed and produced.

Increased memory now permits more thorough error-checking using modular utility subroutines. This contributes to effective courseware and the general productivity of the developer. The later example of inserting a call to a subroutine named ClrKeys before presenting a screen prompt avoids a common problem in which accumulated key presses are returned at inopportune times. Also, a generalized input subroutine illustrates the mechanism for passing arguments between a main routine and the contents of a routine within the user's library.

6.1 Technical suggestions

In very general terms an effective teaching simulation is one in which the student feels in control, knows just what to do next, and sees an immediate response after each input. The following programming suggestions may seem like platitudes but they easily can be overlooked when the developer's attention is upon physiological validity or program logic.

6.1.1 *Consistent operating procedures*

Within any one program it is imperative to have the same control keys perform analogous functions, not a "+" at one point, an "I" at another, or an "up arrow" to denote an incremental increase in some quantity. The labels on some of the PC keys carry labels which may be associated with certain functions. The Escape key often is used to exit, cancel, or abort an operation. It is helpful to use the first letter of a command name, such as "Q" for Quit.

The function key labels, F1 – F10, are not descriptive. When extensive lists of cryptic commands are required one should have these in an easily accessible menu which can be called by a well-defined key, such as "M" for Menu or "H" for Help. These accessory aids may be a window temporarily superimposed over the current simulation. Or, if using the text mode (SCREEN 0) with CGA, instantly viewed or replaced from an alternative screen page.

Consistency also applies to placing specific material at the same point on different displays, and with consistent colour if that is being used. That is, the prompt for a command should not be sometimes at the top and other times at the bottom of the screen.

6.1.2 *Errors*

The logical complexity of large multiple-variable programs allows an astronomical number of possible combinations which can not be checked thoroughly except through classroom experience. At the programming level there are both preventative and corrective steps which can be taken through use of generalized subroutines. Even then a significant portion of program development time is spent in trying to make it bullet-proof.

Section 6.2 details specific software which avoids wrong entries from the keyboard. The PC keyboard buffer easily accumulates extraneous key presses. In order to clear

Simulation techniques for teaching

this buffer a call to a subroutine should precede every issuance of a prompt calling for user keyboard input.

The default initial state for the case of keyboard letters and the keypad keys is not always the same for all computers. Many a program has been written which checks only for upper case letter responses which are ignored on hardware set for lower case. *Listing 1* contains routines for setting case and keypad keys by simply calling appropriately named subroutines, for example Caps or Cursors.

The simple keyboard input command INPUT X is vulnerable to erroneous input. *Listing 2*, described in a later section, is a general-purpose input subroutine which ignores nonnumerical entries, checks against reasonable minimum and maximum values, and also can return the ASCII code for single key entries including the Escape key.

When BASIC programs encounter an error during execution they generate a numerical code associated with the cause. Large program space allows the luxury of letting the program handle the error gracefully without a need for aborting, as is so often the case for interpreter software. As a specific example, if a file of initial conditions for a simulation model can not be found on the disk, it possible to have the program generate these conditions using default settings within the model itself.

6.1.3 *Responsiveness*

It is frustrating for an operator to press a key and see no visible response even though the program may be going through a lengthy calculation. Thus it is necessary to have a visual clue that changes the screen in some way, even something as simple as a PLEASE WAIT ... message. When this is lacking the natural tendency is to keep pressing the same key, storing information in the keyboard buffer which may have unforeseeable consequences later.

Data entries require a final ENTER key as a terminator while most menu responses can be chosen by single letters. For consistency both numerical and single-key entries both might be terminated with the ENTER key. My own prejudice is that single-letter responses without a terminator are much snappier and easier to use in sequence. If the operator reflexly does include the ENTER after a single letter this can be cleared by a CALL ClrKeys before subsequent input.

Single-letter responses are effected by use of the INKEY= command, as in

```
CALL ClrKeys                              'clear keyboard buffer
LOCATE 10,10: PRINT "Press key";
Ch$ = ""
WHILE LEN(Ch$) <> 1: Ch$=INKEY$: WEND
```

The program runs in a loop until a key is pressed and transferred to the string character Ch$. In *Listing 1* the latter two lines are put into a utility subroutine called GetChar. Alternatively, the operation could be a user function which returns a value FNCh$.

Computation speed is an important facet of responsiveness. Operations involving operator input are limited more by human responses than by hardware or software. However, time spent in error checking or extraneous activities during a repeating computation loop needlessly slows responsiveness. When the numerical coprocessor speeds computations the display of values in numerical form may limit overall execution speed. If graphic displays are used to observe general dynamic trends speed is improved if the corresponding numerical values are available only as an option.

6.1.4 *Pictorial presentations*

Appreciation of numerical simulations can be improved markedly by combining these with some anatomical schematic. As one example, during simulations of cardiac mechanics the size of an outline of the ventricles can be correlated with the preload and end-diastolic volume. In another example, concentrations from simulations of the countercurrent multiplier in the kidney can be superimposed on an outline of a simple nephron.

This touches upon one of the more exciting aspects of future teaching simulations in which new graphics capabilities will be used for realistic animations having a quantitative rather than an artistic basis. Programs for the first wave of micros are relatively austere and appeal more to instructors than to students. The next phase is going to take video games as their model of presentation as they hold attention by movement, graphics and colour. The mathematical rigor will be used, but hidden from what meets the eye.

6.2 Utility subroutines

This section illustrates the utility of having generalized subroutines which can be called by a name. I have found that having these readily available in a library encourages their use. These are presented merely as examples of what can be done and by no means exhaust the possibilities. Refer to their respective listings to see the functional clarity offered by the structured BASIC.

6.2.1 *Keyboard subroutines*

Listing 1 (see Appendix) contains several simple, illustrative subroutines given descriptive names. As has been mentioned several times the PC keyboard buffer may accumulate extraneous key pressings which confuse future requests for operator input. Before giving the prompt for such input a call to ClrKeys will read repeatedly the INKEY$ function until it contains a null character.

The byte at absolute location 1047 in the PC RAM is coded regarding certain keyboard functions. The state of two of its bits determine whether letters are upper or lower case and whether numbers or cursors are read from the keypad. The listed routine Cursor turns off one bit to force cursor function. The subroutine Caps turns on another bit to force keyboard letters to be capitals.

6.2.2 *Generalized input subroutine*

The simplest BASIC command for receiving information from an operator is the INPUT X command. It is vulnerable for teaching applications where users are not apt to respect the rigid rules it requires, which include accepting only digital characters.

Persons writing interactive software to be used by large numbers of people generally have their own personal generalized input subroutines. Many articles and even commercial products have presented 'bullet-proof' input routines which avoid many common pitfalls and provide conveniences not found in the limited INPUT X command.

Listing 2 shows one generalized input routine. It is presented as an illustration of the structural features allowed by BASIC compilers and as a general guide for users who wish to design their own.

Instead of an INPUT X statement the one used would be

    CALL Entry(X,Min,Max,Fld%,Code%)   'generic case'
    CALL Entry(X,0,100,5%,Code%)   'specific numbers'

where the CALL would go to a library and execute the subroutine named Entry. Note that there are five arguments which are passed and shared with the subroutine. All other variables within the subroutine are local and independent of the main calling program.

The general idea is to return with the variable X set to the value entered by the operator. If the operator enters a value outside the range Min and Max the closer value will be used instead. Fld% is an integer variable which indicates the field size, or number of digits which can be entered and seen on the screen. This prevents excess entries from overflowing into other portions of the screen.

If the first letter is not a digit, that is if it is a letter or the Escape key, the subroutine returns with the value of Code% set to the ASCII value of the pressed key. For example, responding with only the Escape key returns the value Code% = 27.

If ENTER is the key pressed, the value of X is left unchanged at its default value assigned before calling the subroutine. Upon exit from the subroutine the screen displays the resulting value of X, either the default or the entered one.

Examine *Listing 2* for examples of structured BASIC. Consider the subroutine as having two functions, the handling of single key entries (e.g. letters, Esc or ENTER) and digital numerical entries. First the screen location of the cursor is saved and a blank line of Fld% characters is displayed. Then the first pressed key is read and serviced. Note that as keys are entered they are characters which are concatenated to form the string variable X$. At exit this is converted to single precision variable X.

If the first key is an ENTER, having ASCII code 13, the default value of X is displayed and retained. If the first key is *not* a numerical digit (or +/−.) that key's ASCII value is placed in variable Code% and the screen field is erased. These multiple-line IF blocks are not possible in interpreter BASIC.

If the first character Ch$ is a digit it is placed as the beginning of string variable X$. New digits are read in as Ch$ and then concatenated with what is already in X$. Non-digit entries, except for ENTER and Backspace are ignored. If the ASCII value of Ch$ is 13, the string is terminated. If it is 8, the last letter of X$ is deleted. The length of the string, XLen%, is checked against the field size Fld%.

When entry is terminated the string X$ is converted to single precision X and compared with Min and Max and changed to either of these extremes if necessary. If Min and Max = 0 then there is no limit for the entry. Upon exit from the subroutine the value of X is displayed at the screen location of the cursor upon entry.

7. GRAPHICS SUBROUTINES

The scaling, labelling and layout of graphics plots of dynamic simulations is a time-consuming task. This section describes the use of two generalized plotting subroutines which can be compiled and called from a library when needed. Such generalized routines are particularly useful in the initial development of a mathematical model when all that is needed is a visualization of the solution. Once other parts of the program are in place the graphics display can be modified to taste.

The PC BASIC interpreters and compilers provide commands which use the graphics capabilities of microcomputers. The VIEW command selects some portion of the screen (in hardware coordinates) for the graphics commands of CLS, PSET(X,Y) and LINE. The WINDOW command sets the displayed range of variables in problem coordinates. In each of the following examples one subroutine is called to draw and label the axes and determine their ranges. During a simulation a second subroutine is called to plot each point, appropriately scaled.

7.1 Subroutine for one graph

Listing 3 contains the instructions for a generalized subroutine which plots variables X and Y onto labelled axes. The total process involves three parts: drawing and scaling the axes, plotting the data, and support subroutines as given there and included from *Listing 1*. The approach is a variation of that presented by Spain (4) for Apple II microcomputers.

For simplicity the plots are restricted to black and white SCREEN 1 resolution, but it is easy to extend the number of options to include colour and other options.

7.1.1 *Drawing the axes*

The main simulation program must supply the subroutine called Axes with those parameters required for drawing, labelling and scaling the axes. Of the several alternative methods for sharing variables between subprograms this example sets the minimum and maximum values of X and Y in the main program and uses the SHARE command in the subroutine. The title of the plot (Ttl$) and the labels for axes (XLab$ and YLab$) are passed as arguments in the subroutine call.

The generic form of the subroutine call is

$$\text{CALL Axes(Ttl\$,XLab\$,YLab\$)}$$

A specific example is given below. This subroutine draws axis lines with tic marks, places a title at the top and labels each axis. Note that the DisVert(1,YLab$) subroutine of *Listing 1* centres that label vertically in column 1 on the screen.

The minimum and maximum values for each axis are displayed along with the centre value. Upon exit from Axes() the plotting area of the screen is scaled according to the given ranges of the two coordinates.

7.1.2 *Plotting points*

During the computational loop a CALL Plot(X,Y,Z%) does the actual plotting according to the current values of X and Y. If the integer variable Z% = 0 then discrete points are plotted using the PSET(X,Y) command. If Z% = 1 then successive coordinates are connected by line segments. Note that for the latter case the first data pair must be plotted as discrete points to provide a starting point for the first segment.

Figure 4 is an example of the plot for a double exponential using the subroutines under discussion. The end of the main program contains instructions for the compiler to include text files from *Listings 1* and *3*. To reduce the overhead for very large programs these could be placed in a user's library, as discussed in Section 5.

Simulation techniques for teaching

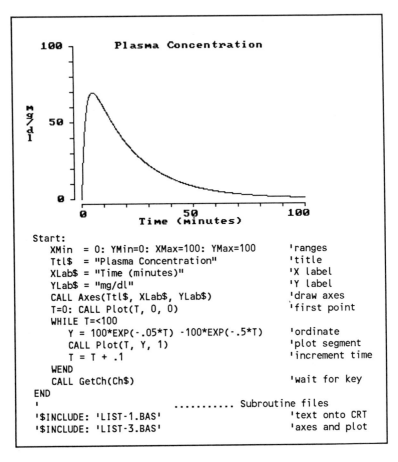

Figure 4. Using the plotting subroutines of *Listing 3*. AXES places title, axis labels, tic marks. PLOT places scaled points or connecting line segments.

7.2 Subroutine for four graphs

The same concept can be extended to plot four different ordinates on four different axes placed on the screen. As before, one subroutine draws and labels the axes while another plots the current data pair on one of the four coordinate systems. In this configuration all four graphs have a common abscissa scale for showing temporal changes in the plotted variables.

Listing 4 shows the two subroutines involved. The main calling program must establish the position of the axes in CRT coordinates in order that they may be shared by the two subroutines. That program also must establish the minimum and maximum ordinates in the arrays YMin() and YMax(), respectively. The abscissa range is from 0 to TMax.

The plotting routine indicates the current coordinates and whether a point or connecting line is to be plotted. In addition it must indicate which of the 4 views are to be used. A simple example for using this subroutine is given in *Listing* 5 and plotted in *Figure* 5. Here the same build up and washout curve is plotted at four different scales.

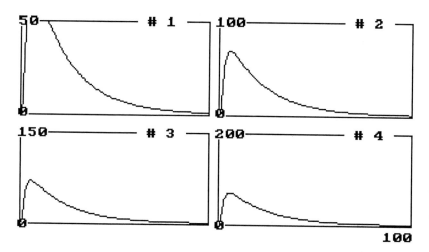

Figure 5. Results of the program in *Listing 5* which plots the same double-exponential function at four different scales using the subroutines of *Listing 4*. AXES4 draws and labels the four sets of axes all of which have the same abscissa. Each ordinate is plotted by calling PLOT4 with coordinates and graph number as arguments.

Notice in *Listing 4* that each call to the plotting routine had to set up the proper viewport and window scales as indicated by the passed argument VW%. Also, the routine must retain the most recent value of coordinates in arrays T0() and Y0() to provide the origin for connecting line segments. There is extensive use of this plotting subroutine in Section 9.

8. USER-DEFINED FUNCTIONS

BASIC programs implement modularity by two kinds of subroutines and by functions. A GOSUB SubA accesses subroutine SubA in which all variables are global. A CALL SubB(X,Y), not usable except for machine code with interpreters, must share variables by some device, such as passing them as arguments in the call statement. BASIC contains a number of mathematical functions, for example square roots and transcendentals. This section illustrates user-defined functions.

The distinguishing feature of a function is that it returns a single numerical value or string. Its advantage over a subroutine is that complex mathematical operations may be replaced with a single expression which starts with the letters FN and contains one or more arguments.

A simple example returns the logarithm of N to the base 10 rather than to the base e as supplied by the BASIC LOG function.

$$\text{DEF FNLG(N)} = (1/\text{LOG}(10)) * \text{LOG}(N)$$

Since functions must be defined before they are used within a program they generally are placed at the start. For PC-BASIC interpreters the definition must be limited to a single line.

The utility of these functions is improved in BASIC compilers which permit the definitions to have several lines. Such flexibility allows for extensive conditional testing and other operations within the function. In contrast to FORTRAN and other languages these functions can not be precompiled and placed in a library. Instead, a given set of functions may be retained in a file and accessed during compiling by the $INCLUDE statement.

Short functions may be defined to perform such utilities as right justifying a string padded with leading spaces, converting an alternate/key combination to the ASCII code for that key, or searching a string array and returning the index of a matched entry. Functions can also be used to perform operations, such as displaying centred lines with underlining or displaying only the first letter of a string at high intensity. These operations are performed by setting the function to some dummy variable, as in Z = FNCenter("Title").

The examples of this section, given in *Listing 6*, provide non-linear functions as required in modeling physiological processes.

8.1 Non-linear function generator

The generalized function generator FNGen(X, X1,Y1, X2,Y2, X3,Y3, X4, Y4) returns a single value of the dependent variable for a given independent variable X. The four coordinate pairs define breakpoints connected by straight line segments.

Values of X less than X1 return the value Y1; values of X greater than X4 return the value Y4. Values which lie between a pair of breakpoints return proportionally interpolated values of Y. Note that the names of the arguments in the function are arbitrary but their order is significant.

Linearized functions may be matched to experimental data instead of explicit mathematical expressions such as polynomials. Computer programs can then plot the entire function or solve one or more iteratively to find solutions for a group of nonlinear relationships. The number of breakpoints may be increased for better fidelity, but all arguments must be supplied to the function. The following two examples approximate ventricular function and venous return curves and then use them in a teaching exercise.

8.1.1 *Ventricular function curve*

The ventricular function curve is used to characterize the heart's output sensitivity to the end-diastolic volume as indicated by the right atrial pressure (RAP). Positive or negative inotropic factors, for example the sympathetics or anesthetics respectively, modulate the expected output for a given RAP. This simplification does not involve numerous other factors, such as afterload, valvular resistance to filling, or filling time.

The definition of a BASIC function may use another previously defined function. The second function in *Listing 6* is an approximation of a ventricular function curve as fit to FNGen. The function FNCO(RAP,INO) has only two arguments: the right atrial pressure (RAP) and an inotropic state multiplier (INO).

Listing 7 contains a program for plotting three ventricular function curves at INO = 0.75, 1.0 and 1.25 times normal. This is illustrated in *Figure 6A*. The program sets up the plotting ranges and calls the Axes subroutine. There are two computational loops. The outer one steps the inotropic multiplier in increments of 0.25 and plots the first data points.

The inner loop steps right atrial pressure and calls the plotting routine with the FNCO(RAP,INO) as the plotted ordinate. At the end the program waits for a key to be pressed before termination.

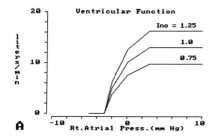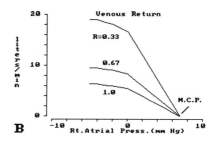

Figure 6. Applications of the nonlinear function generators of *Listing 6*. **(A)** Plots of ventricular function curves for three values of inotropic state (see *Listing 7*). **(B)** Plots of venous return curves for three values of vascular resistance. M.C.P. is mean circulatory pressure (see *Listing 8*).

8.1.2 Venous return curve

Dr Arthur Guyton has popularized the term 'venous return' to indicate the importance of vascular properties in establishing the blood returning to the heart and thus the end-diastolic-volume. His textbooks contain plots of venous return for different right atrial pressures with an abscissa intercept (zero venous return) at a pressure called the mean circulatory pressure.

The horizontal position of the venous return curve is influenced by blood volume and vascular compliance, often combined into the term mean circulatory pressure (MCP). The slope of the descending portion of the curve is influenced by the systemic resistance. According to this scheme during exercise the venous return at a given RAP is increased by the influences on vascular compliance and the reduced resistance.

A common teaching purpose for formulating the venous return is to be able to superimpose its plot on the same axes as the ventricular function curve. The third entry in *Listing 6* is an empirical approximation to a venous return curve with right atrial pressure as the independent variable: FNVR(RAP,MCP,R). Two parameters are included in the argument list. MCP is the mean circulatory pressure, a measure of total blood volume and vascular capacity. The other parameter is the vascular resistance, a factor reduced during exercise. The minimum value of R is forced to be 0.3 to avoid division by zero.

Figure 6B contains three return curves for vascular resistances of 0.33, 0.66 and one times normal. Mean circulatory pressure is the limiting vascular pressure when the heart is pumping no blood. Its value is typically 7 mmHg, as used in the figure. The whole curve is shifted to the left by haemorrhage and to the right by exercise. The first part of the program (*Listing 8*) draws the axes. The outer computing loop steps resistance in increments of 0.33 times normal. The inner one steps RAP from −5 to the MCP and plots the venous return using the FNVR function.

Note that if real numerical values are used as subroutine arguments, such as the value 7. for MCP in this example, it is essential to use the decimal point. This is easy to overlook because BASIC default values are otherwise taken as real numbers when there is no terminating decimal point.

8.2 A simple teaching example

The Guyton presentations superimpose given ventricular function and venous return curves on the same axes. Their point of intersection, a solution for the two non-linear

Simulation techniques for teaching

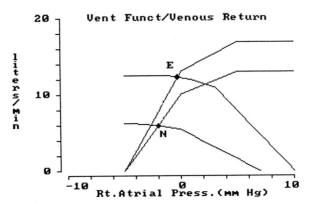

Figure 7. Superimposing ventricular function and venous return curves to find their points of intersection. Point N represents the cardiac output and right atrial pressure for normal ventricular and vascular conditions. The increase of output for point E, simulating exercise, occurs with increased ventricular function combined with decreases in both vascular resistance and vascular capacitance. A barebones program is given in *Listing 9*.

relationships, indicates the resulting right atrial pressure and cardiac output for given mean circulatory pressure and inotropic state. This steady state point tells what RAP is required to give a cardiac output and venous return consistent with producing that RAP.

Cardiovascular teaching often gives logic to the changes in cardiac output and venous/pulmonary pressures in terms of shifts of these two curves and their intersection. A computer program is given in *Listing 9* which combines the numerous modules discussed up to this point. Many desirable refinements have been eliminated in the interest of brevity and simplicity. The point of interest is to see a practical application of the subprograms and to note the clarity possible with structures and named routines.

The central part of the program consists of a series of GOSUB statements which use subroutines to perform specific tasks. The advantage of the GOSUB/RETURN over the CALL subroutine is that the former allows all variables to be global. The program asks the operator to supply three parameters for each simulation.

Figure 7 shows the result for normal parameters and for changes simulating exercise where ventricular function is increased and vascular resistance and capacity are reduced. The net effect is an increased cardiac output at a slightly higher right atrial pressure.

9. LARGE MATHEMATICAL MODELS

The essence of physiology lies in the numerous homeostatic interrelationships involving the body's several organ systems. Microcomputers now have sufficient capacity and power to allow students to explore these regulatory processes by simulating laboratory experiments or clinical pathology and therapeutics. The newer BASIC compilers provide instructors with access to integrative models of a size and complexity once only available to a few specialists.

This section provides a number of examples of the convenience of accessing a large integrated model using combinations of the utility subroutines presented in *Listings 1−8* and described in previous sections. The BASIC source code for the model, consisting of 3000 lines of code in the file MODELSUB.BAS, is too large to be listed here. In addition, three other large text files are required: FUNCTION.BAS, a group of physiological functions used by the model; COMSHAR.BAS, the COMMON SHARE statements with model variables in order; and NAMES.ASC, an ordered list of the names in COMMON. These four files may be obtained on a floppy disc from the author.

Because of its size it is necessary to slight the details and physiological limitations of the model. The listed examples which call the model subroutine are not complete, bullet-proof exercises. The intent is to encourage, by simplification, the use of the model for innovative teaching and to promote simulation as a means of documenting quantitative physiology.

9.1 The comprehensive model

In 1979 Coleman (5) described the comprehensive model HUMAN which combines several details of cardiovascular function with other organ systems, including respiratory and renal functions, and the balancing of heat production with heat loss for temperature regulation. The model keeps track of about 200 computed variables following perturbations of one or more of about 50 parameters. The latter include environmental factors, such as ambient pressure, temperature and humidity; physiological properties, such as heart strength, pulmonary surface area, and renal failure; and various therapeutics, such as selected drugs, dialysis, transfusions, or artificial respiration.

The teaching applications of its simulations include lecture and discussion demonstrations, simulated laboratory experiments, and the diagnosis and treatment of simulated clinical pathology (within limits).

HUMAN originated in a mainframe environment and was written in FORTRAN using a precompiler designed for simulation. It was subsequently repackaged for microcomputers (6) and copies have been supplied to several hundred teaching institutions. More recently it has been recoded in QuickBASIC in a form to take advantage of microcomputer capabilities, including graphic displays.

9.1.1 *Structure of HUMAN*

This section contains only those details of the HUMAN program required to understand the balance of the chapter. The heart of the program is the mathematical model itself, a collection of 3000 lines of commented source code, divided into 30 modules each of which formulate some specific organ function (see *Table 3*). For example, one module

Table 3. Functional modules in the HUMAN integrated systems model.

Pharmacology section	24 h urine collection
Cardiovascular reflexes	Haemodialysis
Cardiac function	Fluid infusion and loss
Heart—reflex interaction	Water balance
General circulation	Sodium balance
Oxygen balance	Acid—base balance
Carbon dioxide balance	Urea balance
Control of ventilation	Potassium balance
Gas exchange	Protein balance
Basic renal hormones	Volume distribution
Status of kidney	Blood volume, RBC mass
Renal excretion	Temperature regulation

keeps track of body temperature by integrating the net sum of heat production and loss. The former is increased by the metabolism of exercise or shivering at low body temperature. Heat loss, through sweat production, influences water and sodium balance calculated in their respective modules. Plasma osmolarity in turn determines ADH which influences the kidney module.

There are currently 479 BASIC real, single-precision variables defined within the model subroutine. These are of four types: physiological or environmental parameters, computed physiological variables, computational parameters and assorted symptom flags or temporary intermediate values. The values of these 479 BASIC variables are kept in memory in a particular sequence by a COMMON statement. There they may be addressed either by name, for example AP for arterial pressure, or as an element in an array, A(20).

The names of several of the computational parameters begin with the letter Z, such as ZDT for integration interval, or ZFLAG6 for the output display interval. Large values of ZFLAG6, as when following chronic changes over a period of days, force several integrative steps between successive displayed points.

The model contains over 60 integrators. In order to set these integrators to their respective initial conditions, as for total body water, heat content, or sodium mass, the parameter ZFLAG is set to 1 and the model subroutine is called one time. Thereafter, for normal computations ZFLAG is set to 0, a condition recognized by the integrator functions. The model also contains 90 non-linear functions similar to those described in Section 8.

The usual sequence is to make one call of the MODEL subroutine to set initial conditions, then set parameters to simulate an experiment, and then progress through a computation/display loop which follows the response of selected variables either in tabular form or on a graph. Extensions of this basic procedure include the ability to save the 479 data array as a disc file for future use, the ability to select which of the 200 computed variables are displayed, and the ability to look at charts of the 200 variables organized according to organ systems.

9.2 A simple example

For the simplest case the math subroutine source code file, here called MODELSUB.BAS, is used as a subroutine in the same compiled module as the program which calls it. *Listing 10* shows such an example. When the compiler reaches the label Model: it reads in two text files, one with a set of physiological functions and the other with the model itself. The subroutine is terminated with a RETURN statement.

The advantage of this GOSUB subroutine is that all named variables within the model are global and are available to the calling routine. This is very convenient during development of the model itself. The disadvantage is that it takes 90 sec to compile this code on a standard 4 MHz PC computer. Once developed and debugged this module may be saved in compiled form for rapid access in the user library.

In the listing the initial steps set integrator initial conditions and all variables to normal values by calling the subroutine once with ZFLAG = 1. ZFLAG6 determines the display interval, 0.25 min here to follow the rapid responses to exercise. In this example the one parameter to be changed is EXER which simulates a level of exercise. It reflects the rate of ATP utilization above resting level and is expressed in the rate of oxygen

Table 4. Output of 3 min simulation of *Listing 10*.

Level of Exercise = ? 1

0.00 min	Press	= 123/85 mmHg		CO =	5350	ml/min
0.25 min	Press	= 131/86 mmHg		CO =	5942	ml/min
0.50 min	Press	= 137/87 mmHg		CO =	6381	ml/min
0.75 min	Press	= 143/88 mmHg		CO =	6915	ml/min
1.00 min	Press	= 147/87 mmHg		CO =	7567	ml/min
1.25 min	Press	= 150/85 mmHg		CO =	8476	ml/min
1.50 min	Press	= 153/83 mmHg		CO =	9451	ml/min
1.75 min	Press	= 154/82 mmHg		CO =	10354	ml/min
2.00 min	Press	= 156/81 mmHg		CO =	11033	ml/min
2.25 min	Press	= 157/80 mmHg		CO =	11630	ml/min
2.50 min	Press	= 158/81 mmHg		CO =	11704	ml/min
2.75 min	Press	= 158/80 mmHg		CO =	12215	ml/min
3.00 min	Press	= 159/81 mmHg		CO =	12285	ml/min

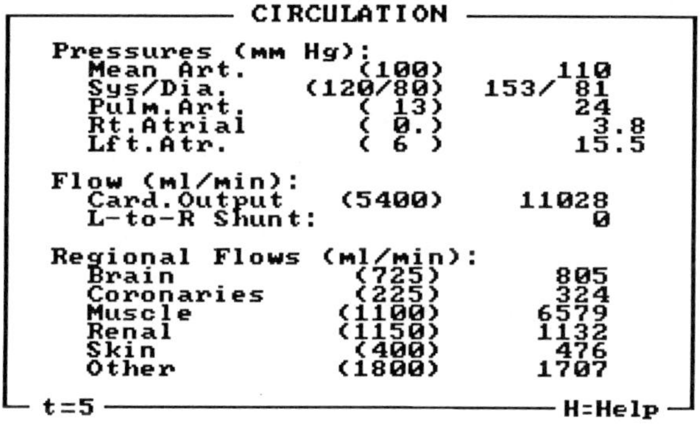

```
┌─────────────────── CIRCULATION ───────────────────┐
│   Pressures (mm Hg):                              │
│      Mean Art.         (100)         110          │
│      Sys/Dia.         (120/80)   153/ 81          │
│      Pulm.Art.         ( 13)          24          │
│      Rt.Atrial         ( 0.)         3.8          │
│      Lft.Atr.          ( 6 )        15.5          │
│                                                   │
│   Flow (ml/min):                                  │
│      Card.Output      (5400)       11028          │
│      L-to-R Shunt:                     0          │
│                                                   │
│   Regional Flows (ml/min):                        │
│      Brain            (725)          805          │
│      Coronaries       (225)          324          │
│      Muscle          (1100)         6579          │
│      Renal           (1150)         1132          │
│      Skin             (400)          476          │
│      Other           (1800)         1707          │
│                                                   │
└── t=5 ──────────────────────────────── H=Help ────┘
```

Figure 8. A chart of several cardiovascular variables as computed by the HUMAN model after 5 min of simulated modest exercise. Normal resting values are within the parentheses. BASIC's PRINT USING command is very helpful in laying out screens of this type.

utilization it would take to supply that ATP aerobically. The indicated value of 1 is in litres of oxygen/min, an amount roughly equivalent to a brisk uphill walk.

The main purpose of this simulation is to show the ease with which any of the computed variables can be accessed using familiar BASIC statements. Here the 3 min simulation calls the subroutine model at 0.25 min intervals. On each of these iterations all modules are updated according to the rates of heat production and loss, water loss, sodium loss, etc. which occurs during each integration interval.

The three variables which are printed are SBP for systolic pressure, DBP for diastolic pressure and CO for cardiac output in ml/min. *Table 4* shows the results of this computation. In practice generally eight such variables are plotted per line in labelled columns. Any or all of the other variables could be similarly examined. *Figure 8*, obtained from another program, shows several of the cardiovascular variables computed in the model after 5 min of simulated exercise. Note particularly the redistribution of cardiac output as compared to the resting conditions given within the parentheses.

9.3 Acid/base balance

Besides following temporal responses HUMAN can also be used to show the interrelationship between two computed variables. This section provides a program to plot the familiar pH−bicarbonate diagram during different kinds of acid and base disturbances. The specific example involves breathing 5% CO_2 with normal or clamped ventilation and with normal or reduced tubular function.

Listing 11 sets up the initial conditions, including the disturbance and respiratory/kidney function, and plots venous bicarbonate versus pH using the PLOT subroutine of *Listing 3*. The subroutine DrawAxes sets up the CRT scales, plots and labels the axes and draws dotted lines representing three pCO_2 isobars and a normal blood−buffer line.

The SetConditions subroutine initializes the model with an integration and display increment of 5 min. There are two options available for establishing pathology. The response to one query determines whether respiration will react to the 5% CO_2 or be fixed. The latter option turns on a simulated artificial respirator (ARTRES=1), sets rate at 12/min and sets tidal volume at 500 ml to establish a fixed ventilation of 6 l/min. In addition, the kidney's ability to conserve bicarbonate can be reduced by setting the tubular function multiplier (EXBB) to the value 0.3 times normal used here. Inspired CO_2 is determined by the parameter FCO2AT, fractional atmospheric CO_2.

During the simulation the model subroutine is called every 5 min and the PH and BICARB variables plotted using a call to the PLOT subroutine. In addition, digital values of several relevant variables are displayed on the screen.

Figure 9 shows the results of simulations for three different conditions. The normal response (A) shows very little departure from resting values of pH and bicarbonate. When the ventilation is clamped at 6 l/min the CO_2 increases to only the 80 mm Hg

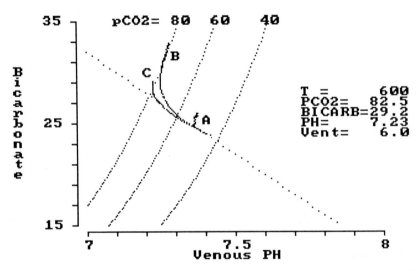

Figure 9. Results of a teaching simulation illustrating three responses to breathing 5% CO_2 for several hours. **(A)** normal response, **(B)** ventilation fixed at 6 litres/min, **(C)** fixed ventilation and reduced tubular function. *Listing 11* contains the program which calls the HUMAN subroutine plus utility subroutines given at the end of this chapter.

isobar maintaining pH by the bicarbonate and protein buffers (B). Note the steady accumulation of bicarbonate with normal renal function.

If ventilation is held constant and tubular function reduced (C) the pCO_2 approaches 80 mmHg but bicarbonate conservation is much reduced so that pH reaches a lower value at the end of the simulation. The model also has the potential for setting hyperventilation and metabolic acidosis for additional stresses on acid/base regulation.

9.4 Multiple graphs

The early HUMAN programs listed computed outputs in columns of a selected eight variables as their values scrolled up the screen on successive iterations of the model subroutine. Myers and Parsons (7) used such information to plot graphs. More recent BASIC versions can plot several variables as a function of time to show general trends.

Figure 10 shows one format using the PLOT4 subroutine of *Listing 4* to plot cardiac output, ventilation, venous oxygen content and plasma lactate over a 15 min simulation at three levels of exercise. The example illustrates the concept of anaerobic threshold. At the two higher levels of 2.5 and 3 litres O_2/min the cardiac output has reached its maximum and the ability of the muscle to extract O_2 is taxed so that glycolysis restores ATP accompanied by an increase of lactate.

Listing 12 contains the program for this simulation. The coordinates and titles for the four graphs are used by the AXES4 subroutine. The initial conditions are set in subroutine SetIC. The four plotted variables carry the BASIC names of COL (cardiac output in l/min), VENT (ventilation in l/min), O_2V (venous oxygen content in cc/ml), and BICARB (plasma bicarbonate in mmol/l).

Note the four separate calls to PLOT4, one for each graph, at each time increment. After satisfying the compute/plot loop control branches back to accept a new value for EXER so that several curves can be superimposed.

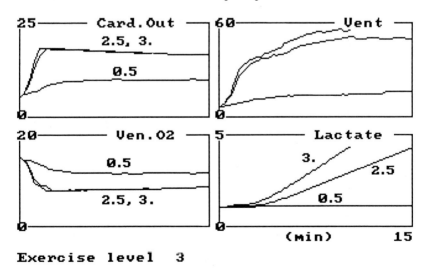

Figure 10. A teaching simulation which illustrates anaerobic threshold. Three different levels of exercise were used. At 0.5 the ATP replacement is achieved aerobically. At levels of 2.5 and 3.0 the supply and extraction of oxygen become limiting so that glycolysis becomes significant. The program, given in *Listing 12*, uses HUMAN and other subroutines.

9.5 Combining model with graphics by text files

Whereas HUMAN keeps track of about 200 computed variables, the screen can plot only a subset of these (four for lecture projections or six for close CRT viewing). For lecture applications the only real difference between a wide variety of possible teaching simulations may be the choice of which few variables are to be plotted. In the context of *Figure 10* that means the names of four BASIC variables, along with the scales and titles for their respective graphs.

This section describes a way to set up the four graphs with any four of the HUMAN variables by using a text editor to create a file with the relevant information in ASCII form. Thus one program, sometimes called a driver, can act to combine the model and plotting subroutines for selected variables without having to compile individual programs. This is particularly useful for preparing dynamic lecture demonstrations in that each presentation consists of a text file containing only 16 ASCII entries.

9.5.1 *Accessing variables by name*

The programs in the previous sections of this chapter display the values of variables by the commands like PRINT AP. There is no way for the operator to enter arbitrary names of variables unless they were part of the PRINT statement when compiled. It is not possible to determine directly the value of a variable which is given only in ASCII form, as from a text file. That is, plotted values must be real variables, not string variables.

The early version of HUMAN, written in FORTRAN, takes advantage of the EQUIVALENCE command in that language. Each of the 479 variables in the model subroutine is assigned a name, such as AP, CO, PH, etc. In addition, the variables are placed in sequence in an array A() in COMMON where they are available to all compiled modules in the program. Through use of EQUIVALENCE the memory location containing the value of arterial pressure carries the labels of both ''AP'' and A(20), etc. for the entire list.

In addition, the names of all variables are placed in ASCII form in an array, called NAMES(), arranged in the same order as the variables are in COMMON. That is, the contents of NAMES(20) has an ASCII AP, etc.

In use, if an operator wishes to know the value of a variable its name is entered in ASCII form at the keyboard. That entry is then compared with each entry in NAMES(). The index for which there is a match is then used to read the corresponding element from the data array A(). As an example, NAMES(123) contains ''EXER'' so A(123) has the value for the exercise parameter.

9.5.2 *QuickBASIC equivalence*

In the present BASIC application the ASCII names of variables to be plotted are read from a text file and compared with the contents of the string array NAMES$(). The index for which there is a match is then used to indicate which entry in the model's data array A() is to be plotted. This means that at some points the program refers to a variable as AP and at other points it accesses A(20), etc.

Although BASIC does not have an EQUIVALENCE statement it is possible for QuickBASIC to share ordered memory locations through use of COMMON in different, separately compiled modules. (I have found no way to achieve this dual naming of memory locations in Turbo BASIC.)

Thus for the present objectives the mathematical model subroutine is moved to a separately compiled module. This has a special advantage of speeding up compiling times for the main loading routine because the large module subroutine can be placed in the user's library. *Listing 13* indicates how the MODELSUB and FUNCTION files have to be combined to be placed in a compiled module.

The first of the modules must contain statements which list the named variables in the order to be placed in COMMON. These statements, of the form COMMON SHARED A, B, C ... have been placed in the text file COMSHAR.BAS and are included at the start of the listing. Note that some dummy variable is placed at the first of COMMON to match the entry A(0) within the main program. The latter is placed there by the BASIC compiler but A(1) contains the first data value.

After being compiled the object file for *Listing 13* is placed in the user's library by the DOS command

c: > BUILDLIB LIST-13.OBJ, SUBS.EXE

This library may be placed in RAM for rapid access during development of the main calling program by calling QuickBASIC from DOS by

c: > QB/L SUBS.EXE

This eliminates the 90 sec delay associated with compiling the GOSUB MODEL ... RETURN sequence in previous listings in this chapter. It also means that entries such as AP or EXER in the MODEL subroutine are available as A(20) and A(123), respectively, in the main calling routine.

9.5.3 *Accessing the data array*

The important qualification for using dual names for variables in QuickBASIC is that the synonymous variable names have to be in COMMON and in separately compiled modules. Thus the main calling program of *Listing 10*, would have to be modified to *Listing 14* which uses the terms A(70) for cardiac output, A(330) for systolic blood pressure, and A(89) for diastolic blood pressure. Obviously, this is inconvenient.

Listing 15 contains a scheme which converts variables named with ASCII strings over to their matching index in the COMMON array A(). The user-defined function FNA("AP") returns the value 20, FNA("EXER") returns 123, etc. for the total data array. Thus PRINT A(FNA("AP")) allows the main calling program to access arterial pressure as calculated in the mathematical model subroutine.

The operations within the listing are described now. The bottom portion transfers the names of variables from the text file NAMES.BAS into the string array NAMES$(). The top portion defines the function FNA(Z$) where Z$ contains the name of the variable to be matched against the entries in NAMES$(). Successive comparisons between Z$ and NAMES$(I%) end when there is a match and I% is returned as the value of the function. If there is no match a 0 is returned to indicate an error condition.

The filling of the NAMES$() array is done once during the initialization of a program. The time required to search for a match for individual variable names depends upon where the variable lies within the array, but averages about 0.1 sec. This is tolerable

Simulation techniques for teaching

for operator iterations but slows computation/plotting loops excessively. For the latter it is less time-consuming to evaluate the indices of the plotted variables only once, saving them as integers for use within the computing loop. This is illustrated by the program in *Listing 16*.

9.5.4 Driver program

The program in *Listing* 16 provides a flexible way of plotting any four variables in HUMAN as indicated by a short ASCII file prepared by a text editor or word processor. Such a file contains four entries for each of the four graphs: the title for the graph, the name of the variable to be plotted, and its maximum and minimum ordinates. A library of different combinations can be accumulated as demonstrations to go with different lectures or discussions which use the model.

The driver program is similar to the one in *Listing 12*, discussed in Section 9.4, except that the graphics parameters are read from a text file rather than being within the compiled program. The additional steps involve opening the proper file, moving its contents into the arrays Ttl=(), YMin(), YMax(), as before. The name of the variable to be plotted is converted to its data array location and saved in Y(). The four sets of axes are plotted and labelled as before. The duration of the plot is entered by the operator.

For flexibility the operator has the option of picking which parameter is to be disturbed and its value. For simplicity this is limited to only one, ambient temperature in a later illustration, but should be able to accept additional ones chosen by the operator. As an example, both ambient temperature and relative humidity might be set. Alternatively, it is possible to expand the size of the text file to include additional information regarding which combination of parameters are relevant for a given simulation. Such limitations are reasonable for student uses.

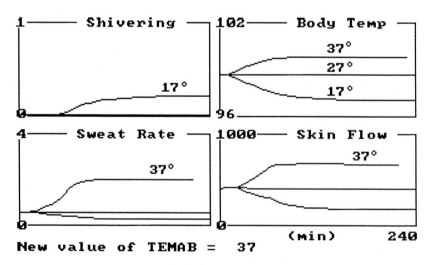

Figure 11. A demonstration of four responses to step changes in ambient temperature. The variables plotted and their scales were read from the ASCII file given in *Table 5* prepared by a text editor. The instructor must enter the names of the parameters to be disturbed and their values. *Listing 16* gives the program which uses the text file to select plotted variables from the HUMAN model.

Table 5. Parameters used for *Figure 11* (from an ASCII file).

Shivering
SHIV
0
1
Body Temp
TEMPF
96
102
Sweat Rate
SWETV
0
4
Skin Flow
SKNFLO
0
1000

Note that because the information required to plot each graph is contained in an array it is possible to call the plot in a FOR loop with the view (VW%) being stepped from 1 to 4. The MODEL subroutine is called only once at each value of time.

Figure 11 contains the plots of three simulations of variables selected by the information given in the text file of *Table 5*. The variables plotted are SHIV (shivering heat production in Cal/min), TEMPF (body temperature), SWETV (sweat value in ml/min), and SKNFLO (skin blood flow in ml/min). The three simulations were run at ambient temperatures of 17, 27 and 37°C. Note the adjustment of heat production and loss and the change in skin flow to adjust thermal conductance.

10. TEACHING PROTOCOLS

Previous sections of this chapter stress the techniques used in programming teaching simulations. But this emphasis should not detract from the broader considerations involving the general objectives to be served, the appropriateness of program contents and the environment in which the materials are to be used. It is not sufficient to simply place courseware in a learning resource centre without providing both motivation and caution for its use. As mentioned before, an enthusiastic teacher provides the link between the technology and the students.

10.1 Objectives

The list of educational objectives which can be served by microcomputers remains open-ended. It started with drill and testing using branching tutorials. The future promises clinical case simulations using artificial intelligence and expert systems. The purpose of the kind of mathematical simulations under discussion here is to provide dynamic answers to the 'what if' question for a finite set of conditions.

In modern parlance such simulations are analogous to electronic spreadsheets which consist of a matrix of cells whose contents are related in some algebraic way. Changing one or more parameters (contents of a spreadsheet cell) of a large model provides simultaneous values for all of the model (other cells), but adds the dimensions of time.

Simulation techniques for teaching

Homeostatic mechanisms, generally speaking, control *amounts* by adjusting *rates* of production or loss. Thus the physiological model includes rates of change within its relationships.

The printed page, out of necessity, presents only a finite number of static cross-sections in the time domain whereas simulation emphasizes the temporal aspects. Mathematical simulations can serve teaching objectives which draw upon the interactions between the organ systems and the fact that the responses to disturbances or pathology take time to develop. For example, the slowness of renal responses to respiratory acidosis correlates with kidney function.

Simulation also presents an opportunity for focusing upon physiology as an integrative discipline during an era dominated by cellular specialization. Introductory physiology courses generally present one organ system at a time supplemented by integrative topics such as exercise, acid-base balance and common clinical pathologies, such as congestive failure. Animal laboratory experiments, important as they are, often become exercises in developing technical skills rather than being lessons in how organ systems interact. Valid simulations provide an opportunity to explore the close coupling of body systems without the limitations of measurement.

10.2 Appropriateness

Experienced teachers are sensitive to the necessity of matching the content of their teaching materials to the background and interests of their students. Whereas engineering students might appreciate a theoretical derivation for turbulence, medical students are more interested in the audible consequences of turbulence at a heart valve. Similarly, mechanical and hydraulic models provide visualization of many basic physiological phenomena. Many of these are being replaced by computer programs.

There is a subtle hazard inherent in using computers models to replace and extend visual models. The limitations of the latter are very obvious so that the range of applicability is not overextended. One does not expect a mechanical model for pulmonary ventilation to compensate for acidosis, but as soon as the model becomes mathematical its basis can be lost to the student. If the instructor is supplied only with the source code it may be impossible to grasp just what is or is not being considered by the model.

Many a program written in interpreted BASIC has been obscured by the necessity to place all conditions of an IF statement on a single line. The structure possible with compiled languages, along with calling procedures by name rather than by line number, helps language-literate instructors get the feel for how a model works, and enables them to make modifications once its structure has been established.

The programmer has at least two options to avoid overextension of the model. One is to limit the input disturbances to within the range for which the model is valid, as for the amount or rate of haemorrhage. These limitations may be built into the program itself or may be inherent in the kinds of questions placed in laboratory or study manuals. As the number of analogue parameters in a non-linear model increases the number of possible combinations can become astronomical so that it is impossible to pretest all of these. One should not underestimate the effort which must be expended.

Instructors also have a responsibility for establishing program validity and relevance before they require students to use it. Materials which were complete as used by their authors may lack critical information when used by others.

By definition *any* model is only an approximation to reality. This should be continually foremost in the minds of users who may be otherwise misled both by the precision of the answers obtained and by extensions for which the model is not valid. At the same time, more advanced students may learn considerable physiology by determining the limitations of a model and incorporating their own improvements. As a specific example, if ventricular strength is increased in HUMAN, to simulate physical training, exercise is terminated rapidly by coronary insufficiency because there is no accompanying increase in coronary vascularity in the model.

10.3 **Environment**

Any given physiological model may be presented in a variety of different formats depending upon the depth of understanding and the environment for which it is to be used.

As an example the Hodgkin and Huxley model for axon excitation may be presented in graphic form, as in *Figure 1A*, during lectures and group demonstrations. Here the objective might be to illustrate the properties of excitation, including refractory period and accommodation. At a more advanced level it may be appropriate to examine how the individual electrochemical driving forces and ionic conductances determine these properties. For even more detail, the conductance factors N, M, and H can be followed, or, single-channel binary changes in conductances can be summed to generate the macroscopic conductances.

My own use of teaching simulations has fallen into three kinds of situations. These have been either in the lecture hall, for individual study or laboratory exercises. The requirements are slightly different for each.

10.3.1 *Group demonstrations*

Simulations which are to be run before a group stress graphic displays over text. The 40-character/line, medium resolution is required for good visibility. Since demonstrations are run by persons familiar with the program operation, there is no need for instructional overhead. The main focus is upon making it convenient to run during an on-going monologue.

Since displayed information comes fast, it is helpful to distribute representative graphs for students' notes. There are a number of memory-resident utilities which will capture screen contents and then reproduce them under control of a word processor. I can get 12 medium-resolution graphics on one side of a sheet of paper.

For small groups the presentation may spark a dialogue if the instructor asks students to predict what will happen for a given set of conditions. It is rewarding to have students think of new situations to be tested.

10.3.2 *Individual worksheets*

Another application I have used is to distribute homework exercises based upon computer simulations using computers of their own or at campus locations. The materials include both a computer disc and individual worksheets which must be returned to the instructor. I scan the latter looking to see that they are grasping the main concepts and using the terms correctly. The topics are generally isolated phenomena, like axon excitation, the arterial windkessel model, or ventricular pressure−volume loops.

Programs and worksheets for this environment need to be both structured and bulletproof. Screens can contain extensive tutorial information. Since the high resolution CRT mode is visible to individuals the screens can contain more information or plot more points for each function.

Things seem to go smoother if the worksheet instructions include representative CRT menus and plotted responses in addition to text material. These printed visual aids help guide the students but are not sufficient to provide either numerical or discussion answers to the questions being asked.

10.3.3 *Laboratory sessions*

In our department the medical/graduate physiology course has a weekly 4 h laboratory session devoted to human studies (respiratory, kidney and exercise), animal experiments and computer simulations. The latter involve integrative concepts covered throughout the previous weeks of the semester.

The simulations tend to mimic traditional laboratory exercises rather than worksheets. This means the screen presentations are not tutorial but simply give the students an opportunity to disturb a model and to record the numerical responses. The thrust is upon taking the data, perhaps plotting it on graphs, but mostly upon the idea of drawing general conclusions about the underlying mechanisms involved.

As stressed above, the range of the disturbances must be matched to those for which the model is valid. It is debatable whether the desire to stretch the model to its limits has any educational benefit or is simply satisfying a sadistic tendency.

One of our laboratory simulation exercises involves acid/base balance in which respiratory and metabolic acidosis/alkalosis is simulated with and without normal lung or kidney function. The other uses HUMAN for experiments of (i) anaerobic threshold, (ii) inspired CO_2, (iii) temperature stress and (iv) haemorrhage. In addition there are several simulated clinical conditions for which the students make a diagnosis and attempt a therapy.

The laboratory environment application may be unstructured, letting the students design an experiment of their own making. One application using HUMAN includes observing the trade-off between oxygen delivery and increased haematocrit with blood doping (8). Those with programming skills may appreciate the opportunity to extend existing models to incorporate new functions.

11. ACKNOWLEDGEMENTS

Although his name is not listed as an author I wish to acknowledge the many contributions of Dr Tom Coleman of the University of Mississippi Medical Center in Jackson. He has generously shared his mainframe model and techniques with the microcomputer community and has sustained my own enthusiasm for placing integrative physiological models into the hands of all physiologists.

12. REFERENCES

1. Randall,J.E. (1987) *Microcomputers and Physiological Simulation*. Raven Press, New York, 2nd edition.
2. Tidball,C.S. and Shelesnyak,M.C. (eds) (1981) *Frontiers in the Teaching of Physiology*. American Physiological Society, Bethesda.
3. Dickinson,C.J. (1977) *A Computer Model of Human Respiration*. University Park Press, Baltimore.
4. Spain,J.D. (1982) *BASIC Microcomputer Models in Biology*. Addison-Wesley Publishing Co., Inc., Reading, MA, p. 345.
5. Coleman,T.G. (1979) *ISA Trans.*, **18**, 65.
6. Coleman,T.G. and Randall,J.E. (1983) *Physiologist*, **26**, 15.
7. Myers,R.S. and Parsons,R. (1985) *Comp. Life Sci. Ed.*, **2**, 25.
8. Holliday,C.W. (1985) *Comp. Life Sci. Ed.*, **2**, 40.

13. NOTE ADDED IN PROOF

Version 4.0 of QuickBASIC (QB4), now available, markedly improves the programming environment over the previous version described in Sections 4 and 5. As source code is entered from ASCII text files or from the keyboard, these are compiled into a p-code which is, in turn, interpreted when the text run is made. Such precompiling checks for syntax and moves most compiling time into the editing session. This, along with improved debugging, rivals the original interpreters for programming convenience.

The one caveat is that QB4 presently supports only the IEEE format for floating point numbers. Internal calculations are more precise but are emulated at a snail's pace on machines that do not have the 8087/80287 maths coprocessors. This is unfortunate because many student work stations eliminate that chip for economy.

Other recent products include a QuickBASIC compiler for the Macintosh and a debugger for Borland products.

Simulation techniques for teaching

APPENDIX
Program listings

Listing 1. Utility subroutines for keyboard and text display.

```
'                               ........ Turn on Caps Lock
SUB CAPS STATIC
    DEF SEG = 0
    Byte% = PEEK(1047)
    Byte% = Byte% OR 64
    POKE 1047,Byte%
END SUB
'                               ........ Key pad for cursors
SUB Cursor STATIC
    DEF SEG = 0
    Byte% = PEEK(1047)
    Byte% = Byte% AND 223
    POKE 1047,Byte%
END SUB
'                               ........ Clear keyboard buffer
SUB ClrKeys STATIC
    WHILE LEN(INKEY$) <> 0: WEND
END SUB
'                               ........ Return key as Ch$
SUB GetCh(Ch$) STATIC
    Ch$ = ""
    WHILE LEN(Ch$)<>1: Ch$ = INKEY$: WEND
END SUB
'                               ........ Rt jusify N @ VT,HT
SUB DisRJ(VT%,HT%,N) STATIC
    N$ = STR$(N)
    IF N=>0 THEN N$=RIGHT$(N$,LEN(N$)-1)
    LOCATE VT%,HT%-LEN(N$)+1
    PRINT N$;
END SUB
'                               ........ Center N on VT, HT
SUB DisCent(VT%,HT%,N) STATIC
    N$ = STR$(N)
    IF N>0 THEN N$=RIGHT$(N$,LEN(N$)-1)
    LOCATE VT%,HT%-LEN(N$)/2
    PRINT N$;
END SUB
'                               ........ Left justify N @ VT,HT
SUB DisLJ(VT%,HT%,N) STATIC
    N$ = STR$(N)
    IF N=>0 THEN N$=RIGHT$(N$,LEN(N$)-1)
    LOCATE VT%,HT%
    PRINT N$;
END SUB
'                               ........ Cent.vert. Z$ at HT
SUB DisVert(HT%,Z$)   STATIC
    FOR I% = 1 TO LEN(Z$)
        LOCATE (10-LEN(Z$)/2)+I%,1èè    PRINT MID$(Z$,I%,1)
    NEXT I%
END SUB
'                               ........ Center Z$ on VT, HT
SUB DisHorz(VT%,HT%,Z$) STATIC

    LOCATE VT%,HT%-LEN(Z$)/2
    PRINT Z$;
END SUB
'                               ........ Right justify @ VT, HT
SUB RJText(VT%,HT%,Z$) STATIC
    LOCATE VT%,HT%-LEN(Z$)
    PRINT Z$;
END SUB
```

Listing 2. Generalized input subroutine.

```
'            Use as: CALL Entry(X, Min, Max, Fld%, Code%)
'When called, waits for keyboard entry at present cursor position
'If first key not +/-/./0-9 then returns ASCII in Code%,
'    thus could use Escape or letters for control codes
'After first digit accepts only digits/./backspace
'Fld% = maximum number characters in field
'Returns single precision entry in X forced within Min/Max range,
'    if Min<>Max; ternminate with ENTER key
'                                   .................................
SUB Entry(X, Min, Max, Fld%, Code%) STATIC
    VT% = CSRLIN: HT% = POS(0)                  'cursor location
    CALL ClrKeys                                'clear key buffer
Start:
    LOCATE VT%,HT%
    PRINT STRING$(Fld%,95);: LOCATE VT%,HT%     'empty field
    Ch$ = ""
    WHILE LEN(Ch$) <> 1: Ch$=INKEY$: WEND       'get character
    ChA% = ASC(Ch$)                             'ASCII value
'                           ............ Service first character
    IF ChA% = 13 THEN                           'ENTRY pressed
        LOCATE VT%,HT%: PRINT SPACE$(Fld%);     'erase line
        LOCATE VT%,HT%: PRINT X;: Code% = ChA%  'display default
        EXIT SUB                                'default return
    END IF
'                           ............ Not #; ASCII into Code%
    IF ChA%<43 OR ChA%>57 OR ChA%=47 OR ChA%=44 THEN  'not digits
        LOCATE VT%,HT%: PRINT SPACE$(Fld%);     'erase line
        Code% = ChA%                            'code return
        EXIT SUB
    END IF
'                           ............ Digital entry
    X$ = Ch$
NewDigit:
    Ch$ = "": XLen% = LEN(X$)
    LOCATE VT%,HT%: PRINT X$; STRING$(Fld%-XLen%,95);  'X____
    WHILE LEN(Ch$) <> 1: Ch$ = INKEY$: WEND     'get char
    ChA% = ASC(Ch$)                             'ASCII
'                           ............ Exit if ENTER key
    IF ChA% = 13 THEN
        LOCATE VT%,HT%: PRINT SPACE$(Fld%);     'clear field
        X = VAL(X$)
        IF Min <> Max THEN                      'no limits
            IF X < Min THEN X = Min             'set within
            IF X > Max THEN X = Max             '  limits
        END IF
        LOCATE VT%,HT%: PRINT X;                'show value
        EXIT SUB                                'value exit
    END IF
'                           ............ Backspace to erase
    IF ChA% = 8 THEN                            'backspace?
        IF XLen% = 1 THEN
            GOTO Start                          'delete field
        ELSE
            X$ = LEFT$(X$,XLen%-1)
            GOTO NewDigit                       'delete char
        END IF
    END IF
'                           ............ Limit field length
    IF XLen% = Fld% THEN
        PRINT CHR$(7);                          'ring bell
        GOTO NewDigit
    END IF
'                           ............ Concatenate character
    IF ChA%<46 OR ChA%>57 OR ChA%=47 THEN       'digits?
        GOTO NewDigit                           'ignore
    END IF
    X$ = X$ + Ch$
    GOTO NewDigit
END SUB
```

Simulation techniques for teaching

Listing 3. Generalized plotting subroutine, illustrated in *Figure 4.*

```
' Use as: CALL AXES("Title","X-Label","Y-Label")
'         CALL PLOT(X,Y,Z%)   Z%=0, point; =1, line segment

'                            Displays title, axes, and their labels
'                            ------------------------------------
SUB Axes(Ttl$,Xlab$,Ylab$) STATIC
   SHARED XMin,XMax,YMin,YMax          'range values shared
   CONST X1=58, Y1=4, X2=298, Y2=164   'corners
   SCREEN 1: CLS:
   CALL DisHorz( 1,23,Ttl$)            'titles, labelsèè  CALL DisHorz(24,23,XLab$)
   CALL DisVert( 1,YLab$)

   CALL DisRJ( 1, 5,YMax)              'display scales
   CALL DisRJ(11, 5,YMin+(YMax-YMin)/2)
   CALL DisRJ(21, 5,YMin)
   CALL DisLJ(23, 8,Xmin)
   CALL DisCent(23,23,XMin+(XMax-XMin)/2)
   CALL DisRJ(23,38,XMax)

   CALL VertTics(11%)                  'ordinate
   CALL HorzTics(11%)                  'abcissa
   VIEW (X1,Y1) - (X2,Y2)              'screen used

   WINDOW (XMin,YMin) - (XMax,YMax)    'variable ranges
END SUB

'                     Plots at X,Y. Z%=0 dots, =1 connecting segments
'                     ----------------------------------------------
SUB Plot(X,Y,Z%) STATIC
   IF Z% = 0 THEN
      PSET (X,Y)                       'single point
   ELSE
      LINE (X0,Y0)-(X,Y)               'line segment
   END IF
   X0=X: Y0=Y                          'save coordinates
END SUB
'                     ...... Local subroutines

SUB VertTics(N%) STATIC                'vert axis, N tics
   LINE (50,4) - (50,164)
   FOR I% = 4 TO 164 STEP (160/(N%-1))
      LINE (46,I%) - (49,I%)
   NEXT I%
END SUB

SUB HorzTics(N%) STATIC                'horiz axes, N tics
   LINE (58,169) - (298,169)
   FOR I% = 58 TO 298 STEP (240/(N%-1))
      LINE (I%,170) - (I%,172)
   NEXT I%
END SUB
```

Listing 4. Subroutine to plot 4 graphs, illustrated in *Figure 5.*

```
'                     ........ Set 4 axes on SCREEN 1
'                     ---------------------
   CONST X1=2,  X2=153,  X3=163, X4=317   'four corners
   CONST Y1=4,  Y2= 76,  Y3=92,  Y4=162   'CRT coordinates
SUB Axes4 STATIC
   SCREEN 1: CLSèè    VIEW (X1,Y1)-(X2,Y2),,3          'view-1
      CALL DisLJ( 1%,1%,YMax(1))
      CALL DisLJ(10%,1%,Ymin(1))
      CALL RJText(1,18,Ttl$(1))
   VIEW (X3,Y1)-(X4,Y2),,3                             'view-2
      CALL DisLJ( 1%,21%,YMax(2))
      CALL DisLJ(10%,21%,Ymin(2))
      CALL RJText(1,39,Ttl$(2))
```

```
    VIEW (X1,Y3)-(X2,Y4),,3                 'view-3
        CALL DisLJ(12%, 1%,YMax(3))
        CALL DisLJ(21%, 1%,Ymin(3))
        CALL RJText(12,18,Ttl$(3))
    VIEW (X3,Y3)-(X4,Y4),,3                 'view-4
        CALL DisLJ(12%,21%,YMax(4))
        CALL DisLJ(21%,21%,Ymin(4))
        CALL RJText(12,39,Ttl$(4))
    CALL DisRJ(22%,40%,TMax)                'max time
END SUB
'
'                   .............. Display Y at time T on VW%
'                                   -------------------------
'       (Z% = 0, plot point; Z% = 1, line segment)
SUB Plot4(T,Y,VW%,Z%) STATIC                'set CRT port
    DIM T0(4), Y0(4)                        'saved coordinates
    IF VW% = 1 THEN VIEW (X1,Y1)-(X2,Y2): GOTO Plt
    IF VW% = 2 THEN VIEW (X3,Y1)-(X4,Y2): GOTO Plt
    IF VW% = 3 THEN VIEW (X1,Y3)-(X2,Y4): GOTO Plt
    IF VW% = 4 THEN VIEW (X3,Y3)-(X4,Y4): GOTO Plt
    EXIT SUB
Plt:
    WINDOW (0,YMin(VW%))-(TMax,YMax(VW%))   'set scales
    IF Z% = 0 THEN
        PSET (T,Y)                          'plot line
    ELSE LINE (T0(VW%),Y0(VW%))-(T,Y)       'plot segment
    END IF
    T0(VW%) = T: Y0(VW%) = Y
END SUB
```

Listing 5. Application of PLOT4 subroutine, shown in *Figure 5.*

```
'           ................. Scales, titles for the 4 variables plotted vs T
DIM SHARED   YMin(4), YMax(4), Ttl$(4), TMax
YMin(1)= 0:  YMin(2)= 0:  YMin(3)= 0: YMin(4)=  0
YMax(1)=50:  YMax(2)=100: YMax(3)=150: YMax(4)=200
Ttl$(1) = " # 1 ":   Ttl$(2) = " # 2 "
Ttl$(3) = " # 3 ":   Ttl$(4) = " # 4 "
TMax = 100
'           ................................ Draw axes, compute & plot
    CALL Axes4                              'draw, label axesèè   T = 0
    WHILE T =< 100
        Y=100*EXP(-.05*T)-100*EXP(-.5*T)    'computed variable
        FOR VW% = 1 TO 4                    '4 viewports
            CALL Plot4(T,Y,VW%,1)           'plot each
        NEXT VW%
        T = T + 1
    WEND
    CALL GetCh(Ch$)                         'wait for key press
    END
                                            'SUBROUTINES
'$INCLUDE: 'LIST-1.BAS'                     'place text on CRT
'$INCLUDE: 'LIST-4.BAS'                     'plot 4 variables
```

Listing 6. Nonlinear function generators used in *Listing 7.*

```
'
'                   ................. Nonlinear function generator
'                                     -------------------------
'             (X=indep.var.; linear segments between 4 breakpoints)
DEF FNGen(X, X1,Y1, X2,Y2, X3,Y3, X4,Y4)
    IF X =< X1 THEN
        FNGen = Y1
        EXIT DEF
    END IF
    IF X =< X2 THEN
        FNGen = Y1 + (X-X1)*(Y2-Y1)/(X2-X1)
        EXIT DEF
    END IF
```

263

Simulation techniques for teaching

```
      IF X =< X3 THEN
         FNGen = Y2 + (X-X2)*(Y3-Y2)/(X3-X2)
         EXIT DEF
      END IF
      IF X =< X4 THEN
         FNGen = Y3 + (X-X3)*(Y4-Y3)/(X4-X3)
         EXIT DEF
      END IF
      FNGen = Y4
   END DEF
   '                       .................. Ventricular function curve
   '                                           -------------------------
   '              (C.O. for Rt. atr. press.; INO = normalized inotropic state)
   DEF FNCO(RAP,INO)
      FNCO = INO * FNGen(RAP, -3,0, -2,5, 0,10, 3,13)
   END DEF
   '                       .................. Venous return curve
   '                                           -------------------
   '            (VR for rt. atr. press.; P=mean circ press; R=resistance to VR)
   DEF FNVR(RAP,P,R)èè    IF R < 0.3 THEN R = 0.3              'minimum resistance
      FNVR = FNGen(RAP, P-11,6.25, P-9,6, P-7,5.5, P,0)/R
   END DEF
```

Listing 7. Program to plot ventricular function curve, shown in *Figure 6A*.

```
'$INCLUDE: 'LIST-6.BAS'              'linearized functions
'                        .......... Axes and labels
XMin = -10: YMin = 0: XMax = 10: YMax = 20
CALL Axes("Ventricular Function"," Rt.Atrial Press.(mm Hg)","liters/min")
'                        .......... Pot 3 intotropic states
FOR Ino = 0.75 TO 1.25 STEP .25      'Ino = inotropic states
   CALL PLOT(-5.,0,0)                'first point
   FOR RAP = -5 to 10
      CALL PLOT(RAP,FNCO(RAP,Ino),1) 'plot ventr. function
   NEXT RAP
NEXT Ino
CALL GetCh(Ch$)                      'wait for key press
END
'                        .......... Utility subroutines

'$include: 'LIST-1.bas'               'place text on CRT
'$include: 'LIST-3.bas'               'axes and plotter
```

Listing 8. Program to plot venous return curves shown in *Figure 6B*.

```
'$INCLUDE: 'LIST-6.BAS'              'functions
'                        ............ Main computation loop
   GOSUB DrawAxes                    'draw, label axes
Another:
   GOSUB GetParam                    'get parameters
   GOSUB DoPlots                     'plot functions
   GOTO Another                      'repeat
'                        ............ Enter parameters at line 25
GetParam:
   LOCATE 25,1: PRINT "Mean Circ Press (";MCP;"?) ";
      CALL Entry(MCP,-4.,10.,4,Z%)
   LOCATE 25,1: PRINT "Vascular Resist.(";R;"?) ";
      CALL Entry(R,0.3,2.,4,Z%)
   LOCATE 25,1: PRINT "Ventr State       (";Ino;"?) ";
      CALL Entry(Ino,.5,2.,3,Z%)
      locate 25,1: print space$(35);:call getch(z$)
RETURN
'                        ............ Draw axes, set scales, defaults
DrawAxes:
   XMin=-10: XMax=10: YMin=0: YMax=20    'rangesèè   Ttl$="Vent Funct/Venous Return
   YLab$="liters/min"
```

```
      XLab$=" Rt.Atrial Press.(mm Hg)"
      CALL AXES(Ttl$,XLab$,YLab$)           'draw axes
      MCP = 7: R = 1: Ino = 1               'default values
   RETURN
   '                                ............ Compute & plot
   DoPlots:
      CALL Plot(-5.,FNVR(-5.,MCP,R),0)      'first point
      FOR RAP = -5 TO MCP
         CALL Plot(RAP,FNVR(RAP,MCP,R),1)   'ven rtrn line
      NEXT RAP
      CALL Plot(-5.,FNCO(-5.,Ino),0)        'first point
      FOR RAP = -5 TO 10 step 5
         CALL Plot(RAP,FNCO(RAP,Ino),1)     'ventr function
      NEXT RAP
   RETURN
   '                                ............ Utilty subroutines
   '$INCLUDE: 'LIST-1.BAS'                  'place text on crt
   '$INCLUDE: 'LIST-2.BAS'                  'input subroutine
   '$INCLUDE: 'LIST-3.BAS'                  'axes and plotter
```

Listing 9. Cardiovascular function curves for exercise, shown in *Figure 7.*

```
   '$INCLUDE: 'LIST-6.BAS'                  'functions
   '                                ............ Main computation loop
      GOSUB DrawAxes                        'draw, label axes
   Another:
      GOSUB GetParam                        'get parameters
      GOSUB DoPlots                         'plot functions
      GOTO Another                          'repeat
   '                                ............ Enter parameters at line 25
   GetParam:
      LOCATE 25,1: PRINT "Mean Circ Press (";MCP;"?) ";
         CALL Entry(MCP,-4.,10.,4,Z%)
      LOCATE 25,1: PRINT "Vascular Resist.(";R;"?) ";
         CALL Entry(R,0.3,2.,4,Z%)
      LOCATE 25,1: PRINT "Ventr State    (";Ino;"?) ";
         CALL Entry(Ino,.5,2.,3,Z%)
      locate 25,1: print space$(35);:call getch(z$)
   RETURN
   '                                ............ Draw axes, set scales, defaults
   DrawAxes:
      XMin=-10: XMax=10: YMin=0: YMax=20    'ranges
      Ttl$="Vent Funct/Venous Return"
      YLab$="liters/min"
      XLab$=" Rt.Atrial Press.(mm Hg)"
      CALL AXES(Ttl$,XLab$,YLab$)           'draw axes
      MCP = 7: R = 1: Ino = 1               'default values
   RETURN
   '                                ............ Compute & plot
   DoPlots:
      CALL Plot(-5.,FNVR(-5.,MCP,R),0)      'first point
      FOR RAP = -5 TO MCP
         CALL Plot(RAP,FNVR(RAP,MCP,R),1)   'ven rtrn line
      NEXT RAP
      CALL Plot(-5.,FNCO(-5.,Ino),0)        'first point
      FOR RAP = -5 TO 10 step 5
         CALL Plot(RAP,FNCO(RAP,Ino),1)     'ventr function
      NEXT RAP
   RETURN
   '                                ............ Utilty subroutines
   '$INCLUDE: 'LIST-1.BAS'                  'place text on crt
   '$INCLUDE: 'LIST-2.BAS'                  'input subroutine
   '$INCLUDE: 'LIST-3.BAS'                  'axes and plotter
```

Simulation techniques for teaching

Listing 10. Simplest use of the HUMAN model subroutine (output in *Table 4*).

```
'                        ............. Set initial normal conditions
DT=0.25: ZFLAG6=DT       'integration/display increment
ZFLAG=1: GOSUB Model     'set integr. init. condx. & all variables
ZFLAG=0                  'future GOSUBS update integrators

INPUT "Level of Exercise = "; Exer

WHILE T =< 3
    PRINT USING "##.## min";T;
    PRINT USING " Press = ###/### mm Hg"; SBP;DBP;

    PRINT USING " CO = #####   ml/min"; CO
    GOSUB MODEL                                    'update variables
    T = T + DT
WEND
END
'                        ............. Subroutine variables are global
Model:
    '$INCLUDE: 'FUNCTION.BAS'    'physiological functions
    '$INCLUDE: 'MODELSUB.BAS'    '3,000 line math model subroutine
RETURN
```

Listing 11. Bicarbonate and pH responses to breathing CO_2, shown in *Figure 9*.

```
'                          ........ pH/bicarbo axes and rangesèè   XMin = 7: XMax =
    GOSUB DrawAxes
'                          ....... New set of parameters
NewValue:
    GOSUB SetConditions                     'condx for run
    WHILE T =< 600
        GOSUB ModelSub                      'update values
        PSET(PH,BICARB)                     'plot point
        LOCATE  8,30: PRINT USING "T =     ####";T;
        LOCATE  9,30: PRINT USING "PCO2=   ##.#";PCO2;   'table
        LOCATE 10,30: PRINT USING "BICARB=##.#";BICARB;
        LOCATE 11,30: PRINT USING "PH=     #.##";PH;
        LOCATE 12,30: PRINT USING "Vent=   ##.#";VENT;
        T = T + 5                                      'increment time
    WEND
    GOTO NewValue
'                          ........ Subroutines:
DrawAxes:
    Ttl$ = "pCO2= 80  60   40         "
    CALL AXES(Ttl$,"Venous PH","Bicarbonate")
    FOR PCO2=40 TO 80 STEP 20
        FOR PH = 7 TO 8 STEP 0.005
            PSET(PH,0.03*PCO2*10^(PH-6.15))             'pCO2 isobars
        NEXT PH
    NEXT PCO2
    FOR PH = 7 TO 8 STEP 0.02
        PSET(PH,32.-(PH-7.)*20.)                        'blood-buffer line
    NEXT PH
RETURN
'                          ......... New parameters
SetConditions:
    ZFLAG=1: ZDT=5: ZFLAG6=5: T=0: GOSUB ModelSub: ZFLAG=0
    CALL Caps
    LOCATE 25,1: INPUT; "Fix Vent"; Z$        'artificial respir?
    IF Z$ = "Y" THEN
        ARTRES = 1: ARRT = 12: ARVOL = 500    'set rate and tid.vol.
    END IF
```

```
        LOCATE 25,1: INPUT; "Renal Fail"; Z$
        IF Z$ = "Y" THEN
            EXBB = 0.3                              'low tubular function
        END IF
        FCO2AT = 0.05                               'inhale 5% CO2
    RETURN
    '                                       ........ Mathematical model
    ModelSub:
        '$INCLUDE: 'FUNCTION.BAS'                   'physiological func
        '$INCLUDE: 'MODELSUB.BAS'                   '3000 line model
    RETURN
    '                                       ......... Utility subroutines
    '$INCLUDE: 'LIST-1.BAS'                         'text subroutines
    '$INCLUDE: 'LIST-3.BAS'                         'plot subroutie
    END
```

Listing 12. Demonstration of anaerobic threshold, shown in *Figure 10.*

```
'                                       ......... Set up 4 graphs
DIM SHARED YMin(4), YMax(4), Ttl$(4), TMax      'used with PLOT4
YMin(1)= 0: YMin(2)= 0: YMin(3)= 0: YMin(4)= 0  'ranges
YMax(1)=25: YMax(2)=60: YMax(3)=20: YMax(4)= 5
Ttl$(1)=" Card.Out " : Ttl$(2)=" Vent "         'titles
Ttl$(3)=" Ven.O2 "   : Ttl$(4)=" Lactate "
TMax = 15
CALL Axes4                                      'draw, label 4 axes
LOCATE 22,28: PRINT "(min)";
'                                       ......... Each exercise value
NewValue:
    GOSUB SetIC                                 'set initial conditions
    LOCATE 24,1: PRINT "Exercise level ";
    CALL Entry(EXER,0.,3.,4,Z%)                 'set exer in ltr O2/min
    CALL PLOT4(T,COL,     1,0)                  'initial points
    CALL PLOT4(T,VENT,    2,0)
    CALL PLOT4(T,100.*O2V,3,0)
    CALL PLOT4(T,BLAC,    4,0)
'                                       ......... Compute/plot loop
    WHILE T = <15
        CALL PLOT4(T,COL,     1,1)              'caradiac out, l/min
        CALL PLOT4(T,VENT,    2,1)              'ventilation, l/min
        CALL PLOT4(T,100.*O2V,3,1)              'venous oxygen cc/dl
        CALL PLOT4(T,BLAC,    4,1)              'blood lactate
        GOSUB ModelSub                          'update all variables
        T = T + DT                              'increment time
    WEND: GOTO NewValue                         'superimpose another
'                                       ......... Subroutines:
SetIC:                                          'initial normal values
    DT    = .25: ZFLAG6 = DT                    'display interval
    ZFLAG = 1: GOSUB ModelSub                   'set model initial condx
    T     = 0: ZFLAG = 0                        'unclamp integrators
RETURN
'                                       ......... Mathematical model
ModelSub:
    '$INCLUDE: 'FUNCTION.BAS'                   'physiological functions

    '$INCLUDE: 'MODELSUB.BAS'                   '3000 line math model
RETURN
'                                       ......... Utility subroutine
'$INCLUDE: 'LIST-1.BAS'                         'place text on screen
'$INCLUDE: 'LIST-2.BAS'                         'general input subroutine
'$INCLUDE: 'LIST-4.BAS'                         'plot 4 variables
END
```

Simulation techniques for teaching

Listing 13. Using HUMAN model as a precompiled subroutine.

```
'     Compile this for user's library during program development,
'     for actual use, LINK with calling program.

COMMON SHARED Dummy                   'variable array(0) not used
'$INCLUDE: 'COMSHAR.BAS'              'COMMON SHARED variables
'$INCLUDE: 'FUNCTION.BAS'             'physiological functions
'                         Note: above are global for compiled module
SUB MODEL STATIC
    '$INCLUDE: 'MODELSUB.BAS'         '3000-line math model
END SUB
```

Listing 14. Accessing HUMAN variables in an array.

```
'Requires user library SUBS.EXE compiled from LIST-13.BAS
' Note awkwardness of using model values as elements in array A()
'
DIM A(479)              'model is an array
COMMON A()              'share with user library
DT    = 0.25            'integration interval
A(8)  = DT              'ZFLAG6, print interval
A(3)  = 1               'ZFLAG, sets model integrators
CALL MODEL              'set initial conditions
A(3)  = 0               'unclamps integrators
T     = 0               'time

Input "Exercise level = "; A(123)            'same as EXER

WHILE T =< 10
    PRINT USING "##.## min"; T;
    PRINT USING " Press = ###/### mm Hg"; A(330);A(89);
    PRINT USING "  CO = ##### ml/min;"; A(70)
    CALL MODEL
    T = T + DT
WEND
END
```

Listing 15. Accessing array variables by name.

```
'Function returns location of Z$ in array of named variables
' placed in same order as they are used in COMMON.
' Variable names read from an ASCII file.

    DIM Names$(Size%)            'array of variable names
DEF FNA(Z$)                      'return entry # of Z$
    WHILE RIGHT$(Z$,1) = " "
        Z$ = MID$(Z$,1,LEN(Z$)-1)
    WEND                         'trim right spaces
    I% = 1
    WHILE (Names$(I%) <> Z$) AND (I% =< Last%)
        I% = I% + 1              'look for match
    WEND
    IF (I% > 0) AND (I% =< Last%) THEN
        FNA = I%
        EXIT DEF
    ELSE
        FNA = 0
    END IF                       'return 0 if no match
END DEF
'               ----------------
'Move names of COMMON variables from file VNAMES.ASC into array
' Names$() in same order as in COMMON.
```

```
OPEN "VNAMES.ASC" FOR INPUT AS #1     'read names into Names$()
I% = 0
WHILE NOT EOF(1)
   I% = I% + 1
   INPUT #1, Names$(I%)
WEND
Last% = I%                            'number of names in file
CLOSE #1
```

Listing 16. Program which uses text file for names and scales of four HUMAN variables to be plotted. Example of *Table 5* is plotted in *Figure 11*.

```
' Requires model in user-library as in Listing 13
' When executed asks for file with names, ranges of variables to plot

CONST Size% = 480              'number of variables in COMMON
DIM A(Size)                    'model variables in an array
DIM SHARED YMin(4), YMax(4), Ttl$(4), Y(4), TMax
COMMON A()

'$INCLUDE: 'LIST-15.BAS'        'access data array by
'                                variable names
GOSUB DispInfo                  'get display information
'                               '   from a text file
CALL Axes4                      'draw, label 4 axes

LOCATE 22,28: PRINT "(min)";
NewValue:
   GOSUB SetIC                  'set initial conditions
   FOR VW% = 1 TO 4
      CALL PLOT4(T,A(Y(VW%)),VW%,0)   'plot initial points
   NEXT VW%
   WHILE T = < TMax             'plot 4 variables
      FOR VW% = 1 TO 4
         CALL PLOT4(T,A(Y(VW%)),VW%,1)  'use line segments
      NEXT VW%
      CALL Model                'update variables
      T = T + A(FNA("ZFLAG6"))  'plot at integr. increments
   WEND
   GOTO NewValue
'                       ........ Subroutines:
DispInfo:
   SCREEN 0: WIDTH 80: CLS
   LOCATE 10,10
   INPUT "Get displayed variables from file "; File$
   OPEN File$ FOR INPUT AS #1
   FOR VW% = 1 TO 4             'each graph
      LINE INPUT #1, TTL$(VW%)  'title
      INPUT #1, Z$: Y(VW%) = FNA(Z$)    'index of variable
      INPUT #1, Z$: YMin(VW%) = VAL(Z$) 'minimum
      INPUT #1, Z$: YMax(VW%) = VAL(Z$) 'maximum
   NEXT VW%
   CLOSE #1
   LOCATE 12,10: INPUT "Plot increment (min)"; DT
   LOCATE 14,10: INPUT "Plot duration  (min)"; TMax
RETURN
'                        ......... Set conditions for run
SetIC:
   T = 0: A(FNA("ZFLAG6")) = DT         'integ/display increment
   A(FNA("ZFLAG")) =1: CALL Model       'set integrators
   A(FNA("ZFLAG")) =0: CALL Caps
   LOCATE 25,1: INPUT; "Change Parameter Named ";Z$: Z% = FNA(Z$)
   LOCATE 25,1: PRINT "Change ";Z$; " from "; A(Z%); " to ";
   CALL Entry(X,0.,0 ,5,Q%); A(Z%) = X
   LOCATE 25,1: PRINT "New value of ";Z$;" = ";A(Z%);SPACE$(10);
RETURN
'                        ......... Utilities subroutines
'$INCLUDE: 'LIST-1.BAS'          'place text on CRT
'$INCLUDE: 'LIST-2.BAS'          'input subroutine
'$INCLUDE: 'LIST-4.BAS'          'plot 4 variables
END
```

APPENDIX

Suppliers of Specialist Items

ACM Algorithms Distribution Service, IMLS Inc. (6th floor), NBC Building, 7500 Bellaire Boulevarde, Houston, TX 77036, USA.
ASYST, Macmillan Software Co, 866 Third Avenue, New York, NY 10022, USA.
Cambridge Electronic Design 1401, Cambridge Electronic Design Ltd, Science Park, Milton Rd, Cambridge CB4 4BH, UK.
Coreco, Incorporated, 555 St Thomas Street, Longueull, Quebec J4H 3A7, Canada.
Dage-MTI Inc.(TV Cameras), 208 Wabash Street, Michigan City, IN 46360, USA.
Data Translation DT2801A, Data Translation Inc., 100 Locke Drive, Marlboro, MA 01752, USA.
Data Translation Ltd, The Business Centre, Wokingham RG11 2QZ, UK.
Digital Precision, 222 The Avenue, London E4 9SE, UK.
Digitimer Ltd, 14 Tewin Court, Welwyn Garden City, Herts AL7 1AF, UK.
Electroplan Ltd, PO Box 19, Orchard Rd, Royston SG8 5HH, UK.
Eutectic Electronics Inc., (Serial section reconstruction system and neuron reconstruction system) 6808 Jersey Court, Raleigh, NC 27612, USA.
Frequency Devices, 25 Locust St, Haverhill, MA 01830, USA. **Lyons Instruments Ltd**, Ware Rd, Hoddesdon EN11 9DX, UK.
Gem Programmers Toolkit, Digital Research Inc., Box DRI, Monterey, CA 93942, USA. **Digital Research (UK) Ltd,** Oxford House, Oxford St, Newbury RG13 1JB, UK.
Hewlett-Packard Corporation (Plotters), 16399 West Bernardo Drive, San Diego, CA 92127, USA.
Hybrid Systems Corporation, 22 Linnell Circle, Billerica, MA 01821, USA.
IBM AT Microcomputers, IBM Corporation, Personal Computer Division, Boca Raton, FL 33432, USA.
Imaging Technology Inc., (TV Digitizer Board), 600 West Cummings Park, Woburn, MA 01801, USA.
Kistler Instrumente AG, CH-8408 Winterthur, Switzerland.
Mostek 1215 W., Crosby Rd, Carrollton, TX 75006, USA. MOSTEK GmbH, Talstrasse 172, D 7024 Filderstadt-1, FRG.
Media Cybernetics, Inc., (Imaging and particle counting software) 8484 Georgia Avenue, Suite 200, Silver Spring, MD 20910, USA.
Mouse Systems Corporation, 2600 San Tomas Expressway, Santa Clara, CA 95051, USA.
Moroz Biomeasurement Systems, 98 Forestgate Drive, Hamilton Ontario, L9C 6A3, Canada.

Appendix

NAG PC50 Software Library. Numerical Algorithms Group, Oxford University, Oxford, UK.

Nascom Microcomputers, Lucas Logic Limited, Welton Road, Wedgnock Industrial Estate, Warwick CV34 5PZ, UK.

NEC Home Electronics Inc.,(Colour Monitors), 1255 Michael Drive, Wood Dale, IL 60191, USA.

Neurolog System. Digitimer Ltd, 14 Tewin Court, Welwyn Garden City AL7 1AF, UK.

Northern Digital, Inc., 403 Albert Street, Waterloo, Ontario, N2L 3V2, Canada.

PCLAMP, Axon instruments Inc., 1437 Rollins Rd, Burlingame, CA 94010, USA.

PolyData Microcentre APS, Strandboulevarden 63, DK-2100, Copenhagen 0, Denmark.

RS Components, PO Box 99, Corby, Northants NN17 9RS, UK.

Summagraphics Corporation, 35 Brentwood Avenue, Fairfield, CT 06430, USA.

Tecmar Labmaster, Scientific Solutions, 6225 Cochran Rd, Solon, OH 44139, USA

Zeiss Microscopes, Carl Zeiss, Inc., One Zeiss Drive, Thornwood, NY 10594, USA.

INDEX

2's complement, 10
10% cosine bell data window, 84, 85
25 Pin D connector, 12
50% threshold, 80
6502, 5
8086, 61
8088, 61
80286, 61, 180
80386, 61

Absolute addressing, 13
Acid/base balance, 250
ADC, see Analogue-to-digital converter
Aliasing, 58
Amiga, 225
Amplitude histogram, 78, 79
Analogue-to-digital converter (ADC), 1, 8, 13, 21, 22, 23, 24, 52, 53, 54, 57, 63, 66, 78, 79, 161, 180, 181, 182, 187
Anatomical calculations, 104–106
Angiocardiography, 174
Apple II, 51, 224
ASCII code, 8–12
ASYST, 91, 92
Assembler, 12, 15, 19, 27, 89
Atari, 225
Autocorrelation function, 83
Automatic measurements, 73–75

BASIC, 1, 2, 20, 24, 54, 90, 92, 221–269
Battery backed RAM, 6
Baud rate, 7
Bessel filter, 58, 84
Bit, 4
Binary,
  image, 130
  numbers, 4, 10
Biofeedback, 211
Bit-plane, 136
Buffer, 11, 12, 67
Butterworth filter, 57, 58, 188
Byte, 5

C, 2, 229
CAT, 130, 175, 205
Camera lucida, 97, 98
Cardiac image handling, 174–176
Catheterization, 174
CCD, 134
CED interface, 53, 92
Cell analysis, 148–154

Centronics interface, 8, 11
CGA, 63, 65, 226, 227
Chaining subroutines, 235
Channel open/close transition detection, 79–82
Characters, 4–8
Chebyshev filter, 58, 84
Chi-squared, 76, 77
Classification of QRS complexes, 170
Clipping, 137
Clock cycles, 7, 13, 23, 61
Closed loop control of locomotion, 198
Clustering, 120
Colour images, 131
Compiled BASIC, 232–237
Computer controlled,
  exercise, 206–208
  movement, 211–218
  walking, 211
Contrast enhancement, 124
Control characters, 12
Conversion rate, 53, 67, see Analogue-to-digital converter
Convolution filter, 161
Counting,
  anatomical features, 101
  nerve spikes, 17
CT, see CAT
CTC (counter/timer circuit), 17, 18, 19, 20, 23
CTS (clear to send), 12
Cubic spline function, 108, 188
Cursor, 97
Curve fitting, 75–77, 168
Cut off frequency, 58

DAC, see Digital-to-analogue converter
Data, 9
  bytes, 4, 22
  storage, 62
  tablet, see Graphics tablet
DCE (data communication equipment), 12
DDT (CP/M), 15
Decimal, 10
Density,
  anatomy, 105
  measurements, 105
Depth clues, 106, 107
Digital,
  filtering, 162
  I/O, 54

273

Index

image manipulation, 129
signal manipulation, 72
Digital-to-analogue converter (DAC),
  8, 16, 17, 19, 20, 25, 52, 86
Digitizing analogue signals, 52
Dilation, 152
Direct memory access, 56
Discs, 7
Display adaptors, 64
DMA, 56, 67, 68
Dot matrix printer, 8, 64, 91, 136, 183
DSR (data set ready), 12
DTE (data terminal equipment), 12
DTR (data terminal ready), 12

ECG,
  aquisition, 160
  data handling, 157–174
  recognition and parameter measurement, 162–173
Edge detection, 143
Educational objectives, 256
EGA, 183, 226, 227
Electrical stimulation bracing system, 213, 214, 215
Electrocardiograph, 160, 161
  patterns, 159
Electrophysiological analysis software, 65
EMG analysis, 21, 22, 182, 188
EPROM, 5
Equilibrium research application, 17–25
Erosion operations, 152
Error prevention, 237, 238
Errors, 20, 77
Exercise,
  aerobic, 204
  anaerobic, 198
  devices involving computer-controlled movement, 198–211
  induced by electrical stimulation, 199
Expansion bus, 61

Fast Fourier transform, 83, 61, 188
Feature,
  measurement, 147, 152
  selection, 149
FES induced exercise, 194
FFT, *see* Fast Fourier transform
File control characters, 12
Filters, 57, 58, 84
Floppy disc, 7, 62
FM tape recorder, 52, 87
FORTRAN, 1, 66, 89, 90, 229

Focus axis readings, 117
Foot ground contacts, 179, 187
Force plate, 187
Form factor, 103
Formatting, 7
Frame,
  of reference, 186
  stores, 124, 136
Freeman chain code, 146
Frequency domain filter, 162
Functional electrical stimulation, 193–198

Gait, 180
GEM, 64
Graphics,
  display, 11, 63
  interfaces, 65, 225, 226
  tablet, 96, 100
  virtual device interface, 90, 91
Graph,
  drawing routines, 241–243, 251, 252
  plotter, 65
  plotting subroutines, 241–243
Grey level, 106, 122, 130, 154
  histogram, 137

Handshake, 11, 12
Hard,
  copy, 64
  disc, 7
Hexadecimal, 10
Hidden line removal, 108
Higher level language, 24
High-pass filter, 58
History of computers in physiology, 1–2
HPGL, 65
HUMAN program, 247–255

IBM,
  model, 25, 225
  PC, 1, 51, 61, 180, 224
Image,
  analysis, 129, 145
  enhancement, 123–125
  manipulation, 129–133
  reconstruction, 131
  restoration, 130
  rotation, 107
  smoothing, 142
  storage, 130
Inspection,
  of records, 69–73
  of transitions, 81
Instruction times, 13, 14

274

Index

Interactive techniques, 236
Interfaces, 2, 17, 53
Interrupts, 15, 20, 23, 56
Ion channel analysis, 77, 78

Joint,
  kinematics, 188
  moments and forces, 189

KERMIT, 3
Kistler force plate, 181, 188

Laboratory interface unit, 52, 53, 54
Laser printer, 65
LCD, 228
Leak current subtraction, 72, 73
Library, 77, 236
Linking compiled programs, 236
LOTUS 1-2-3, 90, 91
Local operators, 141–145
Locomotion,
  data analyses, 183–191
  laboratory equipment, 180–183
  temporal analysis, 188
Logo, 229
Lorentzian curves, 85–86
Low-pass filter, 58, 79

MacAid, 236
MacDope, 236
Machine code, 4–17, 12, 14, 20, 33
Macintosh, 224
MacMan, 236
MacPee, 236
MacPuf, 236
Magnetic,
  discs, 62
  media, 6
Marker kinematics, 179
Marking cell locations, 100
Mathematical models, 247–255
Maths coprocessor, 51, 227, 61
MEPC, 75
MEPP, 70
Membrane potential, 51
Memory, 4
  address, 5
  type, 5
Merging section profiles, 115
Microdensitometer, 135
Microprocessor, 12
Mnemonic (machine code), 13, 115
Monitor, 136
Morphological transforms, 152
Motion analysis, 182

WATSMART, 187
Mouse, 99
MS-DOS operating system, 61, 89, 224
Multiplexer, 22, 54
Multivariate statistics, 174
Muscle,
  analysis in crab, 21–23
  forces, 189

Nascom microcomputer, 2
Negative numbers, 9
Neighbourhoods, 132
Neurolog system, 18, 61
Neuron,
  parameters, 119, 120
  reconstruction, 112
  structure, 119
NLQ, 64
NMR, 130
Noise, 59
  analysis, 83–86
Non-linear function generator, 244
Numerical coprocessor, *see* Maths coprocessor
Nyquist therorem, 58

Open loop control of locomotion, 197
Orientation, 103, 104

Pascal, 2, 20, 21, 24, 25, 47, 90, 229
Parallel data transfer, 7
Parity, 11
Patch clamp, 51, 66, 77
  analysis, 66, 77–86
PCM (pulse code modulation) recording, 88
pdf (probability density function), 82
PDP 11, 1, 51
Peak,
  amplitude measurement, 75
  complex morphology, 170
  recognition (ECG), 162, 163
Physiological simulation, 221
PIO, 8, 17, 19
Pilot, 229
Pixel, 63, 132
Plotter, 183
Point transformations, 139
Power,
  spectral analysis, *see* Spectral analysis
  spectrum, 84
Pre-processing, 149
Printer buffer, 12
Printers, 8
Programmable clock, 52

275

Index

Programming languages, 89−90
Pseudo colouring, 125
Pulse code modulation recording, 88

QB (QuickBASIC), 223
QRS complex, 169, 170

RAM, 5, 52
Random number, 5
Raster, 106
Refraction problems during reconstruction, 117
Rehabilitation engineering, 193
Relative addressing, 13
Relocatable code, 13
ROM, 5
Rotation, 106
RS232, 12
RTS (request to send), 12

ScoP (simulation control program), 229
Section alignment, 111
Sector, 7
Segmentation, 149
Segment,
  anthropometry, 179, 186
  energy, 189
  kinematics, 188
  momenta, 189
Semiautomatic anatomical data collection, 95−100
Serial section,
  alignment, 111
  reconstruction, 110−120
  three-dimensional measurements, 112
Serial transfer, 3, 6
Servomotor, 17, 21
Signal,
  averaging, 70, 71
  conditioning, 57−61
  ECG, 161
Simulating,
  acid/base balance, 250, 251
  cardiovascular variables, 249, 250
Simulation software, 229
Sinclair QL, 17
Single channel analysis, 77−78
Sobel operator, 143, 144
Software,
  design, 89
  for analysing electrophysiological data, 1−93, 65−88, 65
Spatial,
  coordinates, 98, 99
  resolution, 131

Spectral analysis, 58, 59, 83
Spike counting with CTC, 17−19
Start bit, 6
Stereo pairs, 108
Stop bit, 6
Strobe, 22
Structural parameters in two dimensions, 102−104
Stylus, 97
Supercharge, 17

Tablet graphics, 96−100
Tape drives, 62
Teaching objectives, 256
Terminator keys, 238
Three-dimensional computer graphics, 106−119
Time course fitting, 81
Timing with CTC, 17−19
Tissue wrinkling, 117
Trapezoidal rule for area, 102, 103
Trigger signals, 60, 61
TTL (transistor−transistor−logic), 60
Turbo BASIC, 232
TV digitizer, 122

UART, 6
User library, 235

Vector, 106
Ventricular function curve, 244−247
Video, 121−125, 134
  digitizer, 182
  measurement, 121
  RAM, 10, 11
Voltage,
  clamp, 51, 66, 86
  pulse generation, 86, 87

Winchester disc, 7, 62

XENIX, 62
XOFF, 12
XON, 12

Z80, 13, 14, 16, 17